QUANTUM

Mark D. Ziska

Second Edition 2025

This work of fiction draws inspiration from real places and features a mix of characters. Some are based on actual individuals from the public domain, reimagined for the purposes of the narrative, while others are entirely fictional creations.

This book is dedicated to the brave men and women of the US Air Force Combat Search and Rescue, living and breathing Guardian Angels.

"An alleged scientific discovery has no merit unless it can be explained to a barmaid."

Ernest Rutherford

"Fill your paper with the breathings of your heart."

William Wordsworth

TABLE OF CONTENTS

PRELUDE

KM Marker 150 Camino de Santiago, French Way Village of Santa Clara, Lugo, 18 Years Ago

There was a silence as she awoke. Bella could feel the morning's coming; the incessant rains had stopped. From her gabled space beneath the roof, the sunlight and birdsong welcomed the day. She did not hear the rain that had been pummeling her little village on the Camino de Santiago. It was so beautiful after the rain. Bella knew it would be a good day.

Her zest for life showed in her appreciation of her surroundings and of their quaint café—the delicious smells of the daily baking, fresh coffee roasting, and the fragrance of the forest welcomed the pilgrims. The café was a passion of her mother, Mary, and her adopted aunt, Bienvenia—Bea for short, and Mary's best friend. Positioned along the Camino, it offered a blend of warmth and welcome as implied by Bienvenia's name meaning "Welcome" in Spanish.

Nestled in the village of Santa Clara, it sat strategically on the pilgrim's path, a mere 150 kilometers from Santiago, and offered a taste of local delicacies produced through Mary's culinary passion and the region's rich harvest. Outside, birds sang, insects buzzed, and people laughed. The colors of the forest, from the deep greens of the high-reaching trees to the multicolored flowers of the forest floor, were a delight to the eyes and almost enchanted the patrons.

Santa Clara was a cluster of farms, cafes, and the estate of Señor Arensico. A little chapel dedicated to Santa Clara with its own natural spring was said to have the sweetest water on the Camino. Over the last thousand years, pilgrims carried with them a scalloped seashell as a symbol of their journey. These shells, one of the only possessions of the early travelers, served as a bowl for collecting water as well as a plate for food, and over time became a symbol of the pilgrim.

Santa Clara, located near O Cebreiro, a stunning mountain village in a region of outstanding natural beauty, sat along the French Way, the predominant track of the Camino in Galicia in northwest Spain. It was a verdant region with an Atlantic coastline hosting the famous Cathedral of Santiago de Compostela, the reputed burial place of the biblical apostle Saint James the Great. The word "Santiago", derived from the Latin name "Jacobus", translated to "James" in English. "De Compostela" indicated the geographical location and context of the city and came from the Latin term "Campus Stellae", meaning "Field of Stars".

The French Way, "The Ancient Way of the Stars", through the north of the Iberian Peninsula, has existed as a pilgrimage for over 1100 years, dating back to the 9th century, to the time of King Alfonso II of Asturias when the remains of Saint James the apostle were first discovered in northern Spain. It was believed that Saint James preached the gospel in present-day Galicia and on his return to Jerusalem was beheaded by King Herod in 44 AD. His followers were said to have transported his remains back to northern Galicia where they lay undiscovered until 813. The discovery of St. James' remains in the 9th century led to the creation of the Santiago de Compostela community and the Camino de Santiago pilgrimage.

Quantum

In 1140, the Codex Calixtinus, the first guidebook of the Camino de Santiago, was published. It detailed the routes and infrastructure leading to Santiago de Compostela and was widely interpreted as the first tourist guidebook. Throughout the Middle Ages, the Camino de Santiago remained as popular to pilgrims as Jerusalem or Rome. It was the preferred route thanks to the early construction of hospitals or modern day hostels, Templar forts, castles to house pilgrims, and the long monarchy history of promotional work of the itinerary to Santiago. However, with the Protestant Reformation taking place in the 16th Century in Northern Europe, its popularity diminished. It wasn't until the 20th Century that the number of pilgrims significantly increased to a level never seen in history. With the greatest historical tradition and international recognition, it was undoubtedly the busiest of all the pilgrimage routes.

As the dew sparkled on the grass, Bella's excitement overwhelmed her. "¡Mamá, Mama! Date prisa, abramos, vienen los peregrinos!" Hurry, let's open. The pilgrims are coming. These were not the first pilgrims to pass this morning as many arose with the sun, but they were the first that Bella saw on this gorgeous day.

Mary answered, "Si mi amada estaremos listos." Yes, of course, my beloved. We will be ready.

"Mama, I am so excited. I just love the pilgrims on such a beautiful day. I love being here with you."

Bella was raised in a family of wealth. Her father, Miguel, inherited the generations-old sustainable lumber business; the family had forested land holdings across Galicia. Bella spent time with her father exploring the picturesque beauty of the country. Her father showed no deference to her gender and raised her as he would a boy. She learned to hunt, fish, navigate by the stars, build shelters, and start fires without matches. What Bella loved most was when her father challenged her on the ski slopes after a fresh winter snow. She learned to ski in the mountains of Galicia but was as much a natural at home as on any mountain. She could ski bumps, glades, trees, steeps, a couloir, powder, or the long schussing runs where she really demonstrated grace. Miguel met Mary on a ski holiday, and he wanted for Bella the same love of nature and deep soul love that they experienced.

Bella possessed exceptional intelligence and quickly bypassed all of her peers in her extended community. Mary and Miguel decided to bring in Bea to help home-school Bella. Bea held a PhD in Languages from Salamanca University and was a perfect companion, teacher, and governess. She was poised and intelligent and loved the Arensico family as her own. As a tutor, she pushed Bella and Bella responded. Over time, Bella adopted Bea as her own aunt.

This was Bella's last full season on the Camino. Given her gifted student status and insatiable appetite for learning, all agreed that it would be in her best interest to attend public preparatory school in Oxford. She would be near Uncle Andrew; she could visit Bentley Abbey on weekends, and Bea would accompany her. She could excel beyond the capabilities of the family teachers. But beneath that prospect lay a quiet fear: leaving her idyllic life on the Camino, her excursions with her father and that which she has known her whole life.

Jack, hardly older than Bella, was on a pilgrimage set as a challenge by his father, known to everyone as "The Colonel." The Colonel, true to his name, demanded strict discipline and adherence to his authority. In his world, orders were to be followed without question, and Jack was expected to be seen and not heard. A tough master in the chain of command, The Colonel believed

Jack's generation was soft, sheltered by video games and devoid of real-life experiences. He was convinced the Camino would shape Jack into a "man."

Jack loved and respected his father but relished the freedom of being beyond his reach. As an extrovert, he thrived on meeting strangers and embracing the spontaneity of the Camino.

Jack had been nudged out of his comfort zone to confront the complexities of life. At 18, he was already a junior at the University of Arizona, having started university two years earlier than his peers. Academically, he excelled, but socially, he often felt out of place, not only his age but burdened by the high expectations of his family of achievers.

Arriving at a quaint café along the Camino, Jack used his best, albeit imperfect, Spanish to order his favorite meal: un bocadillo de jamón y manchego, un media luna de arándanos, y un refresco de naranja. A ham and cheese baguette, a blueberry croissant as a treat, and orange soda—it was his comfort food during the pilgrimage.

He settled at a table, savoring the moment. The mingling scents of the forest and fresh bread, the vibrant colors of wildflowers, and the symphony of spoken languages created a sense of serenity. This café felt idyllic, a place he could imagine returning to long after his journey ended.

Bella approached with his order, her steps light yet deliberate. She greeted him with the traditional "Buen Camino."

Jack looked up, his warm smile meeting her gaze. "Buen Camino. ¿Cómo está?"

It was then that lightning struck—figuratively. Bella froze, momentarily captivated by the intensity of Jack's green-hazel eyes. She had served countless pilgrims, yet none had ever left her speechless like this.

Catching her breath, she stammered, "Oh, y-you must be American. I can always tell by the accent. Como te llamas? What's your name?"

Jack grinned. "Hola, Mi nombre es Jack. I'm Jack and yes, I am American, but I'm practicing my Spanish. ¿Cual es tu nombre?¿Es este el café de tu familia? I think that's, 'What's your name and is this your family's café?'"

Bella nodded, her thoughts still a jumble. "Hola Jack, mi nombre es Bella. Yes, it is my family's cafe. My mother and aunt own it. I've grown up here." She paused, working to compose herself. "Let's speak English—it's easier to communicate."

She hesitated, searching for the right words. "This morning, I was so relieved when the rain stopped and the sun came out. It's funny how in English you say, 'the sun comes out.' Bea—my aunt—taught me that, along with phrases in French and Italian."

"Wow! That's amazing," Jack said, clearly impressed. "You're so young and already multilingual."

"I'm not that young! I'm fourteen. What are you, sixteen?" she teased.

Jack laughed, shaking his head. "Not quite. I'm eighteen and a junior at the University of Arizona, in the United States."

Bella raised an eyebrow, her skepticism evident. "No way. I don't believe it."

"It's true," Jack insisted. "I'm studying computer science."

"You don't look any older than me," Bella remarked playfully. "But I'll take your word for it… for now."

Jack chuckled. "Thanks, I guess. Maybe one day looking young will be an advantage. What about you? Any big plans?"

Bella's face lit up. "I want to be a physicist. I'm fascinated by atoms and the forces that govern everything about us. I'm preparing for exams so I can apply to Oxford."

"That's incredible," Jack said. "You seem so confident—I admire that."

Bella hesitated, her smile faltering slightly. "I try to be. But the idea of leaving everything I know behind is… overwhelming. I'll miss the simplicity of life here—the forest, my family, my dog. Sometimes, I wish I could just stay in this moment a little longer."

Jack nodded thoughtfully. "I get it, Bella. You're not alone in feeling that way. The unknown is scary, but I think you're going to do amazing things."

Bella shivered slightly, as if sensing something beyond the moment. "You've given me an experience I can't describe in English, pero in Spanish… una premonición sobre el futuro que nos espera de grandeza." She paused, translating. "It means we're meant to cross paths again."

Another chill raced down her spine, and she shivered again.

"Are you alright?" Jack asked, concerned. "You shivered. I was worried." His emotion took over.

Bella sensed Jack's caring, his emotion. She smiled, her confidence returning. "I'm fine. It's funny—I felt a chill, but now

I feel more certain about the future. Thank you." She stood a little taller, her presence self-assured, as if a quiet transformation had taken place.

Jack's gaze softened. They both chuckled, the moment charged with an unspoken understanding. Two old souls, connecting on a journey through life.

"Well, Jack," Bella said, her voice steady, "enjoy your bocadillo and your Camino. Maybe we'll meet again."

"Buen Camino!" they said in unison, their smiles lingering as if sealing a pact with fate.

PART ONE

CHAPTER 1
HER FAMILY

Bella's Family The Background
Oxfordshire, England and Santa Clara, Lugo, Spain
Formative Years

The Arensico family lived in the mountains of Spain, surrounded by forests that they had sustainably harvested for centuries. They cared deeply for the land, ensuring that each generation would inherit an abundance of carefully planted and responsibly harvested trees for lumber.

Mary, Bella's mother, Lady Mary of Bentley, was born in the United Kingdom as the firstborn child of a very old aristocratic family. Her childhood home, Bentley Hall, with its majestic castle and sprawling grounds, was nestled in the Cotswolds, just outside of Oxford. Mary grew up amidst the privileged estate class, where tradition and heritage were paramount. As the firstborn, her father bestowed upon her the title and the estate, as was his prerogative. Thus, Mary became both Lady and Lord of the manor.

Andrew, her younger brother, bore no animosity, resentment, or jealousy toward Mary. While many aristocratic men might have felt slighted at being overlooked for such honors,

Andrew, Lord Bentley, was different. He had always admired his elder sister for her grace, intelligence, and boundless affection. Their bond was unshakable, forged through mutual respect and a shared understanding of their aristocratic world.

Mary proved to be exceptionally intelligent, earning high marks during her preparatory years and excelling in her studies of History and Modern Languages at Oxford University. Her time at Bentley Hall and Oxford was idyllic, shaping her into a woman of both intellect and compassion.

It was during a ski trip to Spain that Mary met Miguel Arensico, a charismatic and passionate man whose family had deep ties to the Spanish mountains. Their connection was instantaneous—a whirlwind romance that quickly led to marriage and, later, the birth of their beloved daughter, Bella. The family settled into a joyful life in the village of Santa Clara, situated along the Camino de Santiago in Spain. For Mary, however, Bentley Hall remained an inseparable part of her identity. She cherished the memories of her childhood home and delighted in sharing its splendor with her family.

Mary frequently took Bella and Bea to visit Bentley Hall, eager to introduce them to the rich history and beauty of the Cotswolds. These trips became cherished traditions, allowing Mary to bridge the two worlds that defined her life: the rolling hills of England and the rugged mountains of Spain. Together, they created a tapestry of love, legacy, and belonging that would endure for generations.

Like the Spanish village of Santa Clara, where Bella grew up, the village of Bentley Abbey traced its history back centuries. Its market town center and church steeple rose high above fertile fields, rolling hills, and sparkling waters. Situated just outside

Oxford, on the edge of the Cotswolds, the village was surrounded by a picturesque landscape of historic honey-colored stone villages, lavender fields, grazing sheep, and meandering streams. The magical views brought smiles to all who visited.

The name "Cotswolds" originated from the local dialect. "Cots" referred to sheep enclosures, while "Wolds" described the rolling hills of the open countryside. Human presence in the region dated back over 3,000 years to the Early Neolithic period, evidenced by ancient stone burial tombs. Later, the Romans left their mark, building roads, bridges, and homes from the honey-colored stone that defined the area's character. Roman sites from the 2nd to 4th centuries could still be found scattered across the countryside.

Bentley Hall estate stood crowned by its castle, a testament to centuries of history and refined craftsmanship. The castle's turrets once served as defensive strongholds, but now they offered panoramic views of the serene countryside. The ivy-covered stone facade, adorned with intricate carvings, exuded a regal charm. Deep beneath the current castle lay the remnants of the original structure, dating back to the tenth century.

A river flowed through the estate, once serving as the primary route of travel for its inhabitants. The surrounding gardens, meticulously designed and cared for, blossomed with vibrant colors and fragrances throughout the seasons. A tranquil lake mirrored the castle's reflection, creating a scene of perfect serenity that invited peaceful contemplation.

Inside the manor, history unfolded through centuries of elegant design. The grand entrance hall welcomed visitors with polished marble floors, their surfaces glowing softly under the light of a crystal chandelier. The ceiling's intricate moldings and the rich tapestries adorning the walls spoke of the estate's noble past. Antique furnishings, fine art, and aged leather added layers of character, while the warmth of a crackling fireplace filled the air.

As the sun set over the Oxfordshire countryside, Bentley Hall was bathed in a golden glow. The estate, with its harmonious blend of natural beauty and architectural splendor, stood as a timeless monument to the grace and elegance of this idyllic corner of England.

Bella was Uncle Andrew's favorite. She cherished her holidays at the Bentley Hall estate, where Andrew always welcomed her warmly, whether during the golden days of summer or the frosty magic of Christmas. Despite his demanding role as the executive director of a family-owned shipping line, which often took him away from home, Andrew always ensured he was available whenever Bella visited. Their bond, filled with shared laughter and heartfelt conversations, added immense joy to Bella's childhood.

Bella adored her trips from her home in Spain to Bentley Hall. She had a special love for her Uncle Andrew, who not only managed the family estate but also oversaw the surrounding fertile lands—fields of lavender and forests teeming with wildlife. Among her favorite activities were visits to the estate's ponds, home to some very old and cunning fish. She and her uncle often tried their luck fishing, though they never succeeded in catching the elusive big ones. The fish would cleverly wriggle free from the barbless hooks. If they had caught one, it would have been strictly "catch-and-release," as both Bella and Andrew respected the elders of the pond.

As a child and later as a young adult, Bella found pure bliss exploring the countryside on horseback with her uncle. These rides transported her to another world, where the intoxicating scent of lavender and the rich aroma of freshly tilled earth filled the air. Along the way, she discovered Roman ruins, wild gardens, and remnants of her forebears' lives.

Bella often lost herself in fantasies as she explored the undeveloped parts of the estate. She imagined lush gardens flourishing where now only whispers of history remained. She dreamt of the couples who once called these beautiful hills home—those who built lives rooted in the land's abundance. She envisioned a romantic existence filled with plentiful harvests, mild weather, and the laughter of children running through the fields. These couples, with their deep connection to the earth, had raised generations on the very lands she now rode across, leaving behind a legacy of love and resilience.

CHAPTER 2
BELLA

Oxford, Oxfordshire United Kingdom
16 Years Ago

Bella had moved to the UK to prepare for the final stages of her lifelong passion for the sciences, particularly physics, which she found most enticing. As a teenager, she split her time between studying and boarding at Rye Saint Anthony Academy in Oxford. Rye Saint Anthony, an all-girls school founded in 1930 on Catholic Christian values, was a perfect fit for an intelligent young woman from the heart of the Camino in Spain. Bella thrived on the challenges presented by her teachers and the school's proximity to Oxford University's many colleges. She completed her Sixth Form with outstanding marks, excelling in science, math, and especially physics. The advanced classes offered by Oxford gave her a taste of the university experience and solidified her academic aspirations.

Though she loved learning, Bella's favorite moments were those spent on weekends and holidays at Bentley Hall. There, she cherished time with her governess, Bea, who, despite being much older, was her best friend. Bea had been a constant presence in Bella's life—first as a nanny, then as a surrogate big sister, and now as a mentor. Bea had moved to England with Bella when she enrolled at Rye Saint Anthony and had become a partner to Bella's Uncle Andrew. The two had known each other since Bella's mother's college days when Mary and Bea spent holidays and summers at Bentley Hall. Bea's intelligence and linguistic skills made her invaluable to Andrew's worldwide charitable endeavors, and her ability to interact with people of all backgrounds helped her thrive as both a confidant and strategist.

Bella's constant companion at Bentley Hall was Emma, a pug given to her for Christmas by Uncle Andrew two years earlier. Loyal and playful, Emma was irresistibly cute with her fawn markings and black snout. The two were inseparable from the moment they met. Emma followed Bella wherever she went, often with a toy caterpillar, almost her own size, dangling from her mouth. Andrew's older pug, Rex, a black-coated senior dog, was equally loyal to him but demonstrated occasional jealousy when Andrew paid attention to Emma. Together, the three adults and two dogs formed a close-knit family. Whether wandering the gardens or lounging by the hearth, they celebrated birthdays, holidays, and successes together, sharing both their burdens and their joys. For Bella, Uncle Andrew and Bea were more than guardians—they were family.

Andrew loved giving meaningful gifts, always knowing just what would bring joy to those he cared about. Bella's latest treasure was a Rolex Cellini Moonphase, the same model Andrew himself wore and one she had long admired. She had often borrowed his watch, fascinated by its display of a radiant blue night sky and the phases of the moon. The rose gold hands, index hour markers, and pointer date around the outer rim on a white dial captivated her. To Bella, it symbolized the elegance and order of the universe.

In time, Bella came to think of herself as having two sets of parents, always stressing her equal love for all. Her joy was their shared joy, and together, they embraced their unconventional family dynamic.

Bella had always been enthralled by the history of Bentley Hall and its surrounding countryside. She was especially fascinated by the many passageways, tunnels, and cellars of the castle. As a child, she and Bea explored every secret panel, passage, and underground chamber, discovering where some of the tunnels ended in beautiful, secluded gardens. "Perfect meeting places for lovers," Bea would tease. Bella's imagination transformed these tunnels into romantic routes for knights, priests, and lovers seeking clandestine access to the castle.

As she grew older, Bella's insatiable interest in history expanded. She, Bea, and Uncle Andrew often pored over volumes of records detailing agricultural production, taxes, and the estate's upkeep. Bella was surprised to learn that, despite the estate's regal appearance, its upkeep had consumed more than its profits for most of its history.

"Uncle Andrew, we've explored the estate's history in such detail," Bella remarked one day. "But I don't understand—why did the family lose money so often?"

Andrew smiled, always enjoying Bella's romanticized perspective. "Bella, I love your idealistic view of history, but life was difficult for most people. This estate's location was brilliant—it sat at a navigable river with tributaries leading to market towns both near and far. The Roman roads were still in good condition, though keeping the forest from overtaking them was a constant challenge. Our little settlement was literally at the crossroads of commerce."

Bea chimed in with a teasing smile, "Bella and I prefer to think of the estate as a place of beauty and romance. The women here spent their days choosing silks and wools for dresses, matching jewelry to their ensembles, and creating tapestries—all while awaiting the gallant knights who would protect and adore them."

Andrew chuckled. "You may have a more romantic notion of knights than history suggests. While they respected the women within the castle walls, they were also the local militia, enforcing the earl's will. Outside these walls, their respect often didn't extend as far."

Bella tilted her head thoughtfully. "Wouldn't a bridge across the river have simplified life?"

"Building a bridge back then was no small task," Bea explained. "Skilled tradesmen were scarce, and even by the late 12th century, Oxford focused on religion, not architecture. Most builders were self-taught."

Andrew nodded. "Still, by 1391, a bridge was constructed. The earl and local guilds shared the cost. Materials came from the estate, while the guild provided expertise and funding. This bridge quadrupled market commerce, and as trade flourished, so did the village."

"That's amazing," Bella said, her cheeks flushing with pride. "Your ancestors really shaped this area."

"They did," Andrew confirmed. "But don't forget, history isn't all romantic. Outlaws, corrupt clergy, rival landowners, and Viking raiders were constant threats."

"Some things never change," Bea added with a wry smile. "Greed and ambition still drive the world."

Andrew tapped his nose knowingly. "That's why the estate began as a castle, with moats, battlements, and high walls. Safety was paramount."

"And ghosts?" Bella teased, her tongue firmly in cheek.

"Every old estate has ghost stories," Andrew said. "Ours are no different, but if they're here, they're family."

Bea laughed. "If there are spirits, they've always been polite to me."

As the conversation continued, Bella noticed the way Bea and Andrew looked at each other—gentle touches, shared glances, and an unmistakable warmth. It was clear to Bella: Bea and Andrew were more than companions. They were a "thing," and their bond brought a quiet, happy stability to the home they all shared.

CHAPTER 3
PAYMAN

The Sound, Plymouth Bay, off the Plymouth Yacht Club
United Kingdom

Payman Mehri, skipper of the 'Falkin, a 33-meter fiberglass keel boat, enjoyed life as a member of an exceptionally wealthy and powerful Iranian family. He was athletic, confident, and possessed intellectual abilities, all wrapped in a handsome appearance. Life couldn't have been better as he soaked in the view of the beautiful Plymouth coast.

He had always been an athlete capable of accomplishing almost any physical goal he attempted, but sailing was his passion. His natural capability on the water came from his ancestors who had navigated the Persian Gulf and the Arabian Sea.

The 'Falkin was one of the most technologically advanced sailing systems in the world, a vessel Payman had personally outfitted to be as close to perfection as possible. This perfection was designed to allow Payman to compete on the international yachting stage, traversing the oceans of the globe. His ambition was to become the world's leading sailor.

Quantum

The Mehri family was one of the wealthiest in Iran. Their business empire dated back hundreds of years in Persian history, and the family itself had its roots in the land. While other family branches became seafaring traders of the family's agricultural products, his branch remained tied to the land and business. The family wealth had grown exponentially, and their holdings had expanded far beyond their original enterprises.

By the twenty-first century, their wealth extended beyond the land and the sea, as they controlled many of Iran's leading banking institutions. This international financial influence, built by the generational patriarch Amir Mehri, had allowed the family to bypass sanctions imposed on Iran and transfer billions of US dollars into Swiss and other offshore secretive financial institutions. This reach and influence had enabled the family to live on the world stage among the wealthiest families.

Since the Iranian Islamic Revolution of 1979, the elite families of Iran included a cadre of clergy, bureaucrats, Pasdaran Islamic Revolutionary Guards, educators, and business leaders. The current generation of Mehri's was influential in all aspects of Iranian life; business operations remained family affairs, often receiving large government loans for ventures secured solely due to the family's recognized status. Political activities also followed family lines. Several of Amir's children, grandchildren, or first cousins joined the Islamic Republican Party, others became part of the Islamic clergy, and some joined the Pasdaran or the armed forces. Successful members were expected to use their influence to assist less successful ones in establishing themselves; inherent nepotism was viewed as a positive cultural value rather than a form of corruption.

Payman, Amir's favorite grandchild—as the elder unabashedly expressed—had shown signs of being a gifted child from an early age. He learned to read, play the piano, and perform complex mathematical computations before beginning formal education. Payman attended the best preparatory schools in Europe and was groomed for a very special role on the international stage of intellectuals and athletes. As the hand-chosen future leader of the family, Payman embraced his role and excelled in his endeavors. He attended Oxford University, where he conducted graduate research in physics, particularly in the intricate world of the atom.

On the sea, the sky expressed a deep Mediterranean blue that sailors had admired for millennia. In the world of color, Pantone, the global standard, had defined it as Pantone 18-4334 TPG—the Color of the Year, Mediterranean Blue—the most pleasing of blues. Payman had been blessed with eyes the color of this Mediterranean sky. These eyes had always been what people first noticed when meeting him. Those Mediterranean blue eyes engendered trust and created a sense of ease in those who came to know him.

PART TWO

CHAPTER 4
OXFORD UNIVERSITY

Bodleian Library, Physics Club University of Oxford, Oxford England

04 August Bella's 18th year of age

Oxford University was considered the epicenter of education in the English-speaking world, or so those at Oxford would have had others believe. It was true that the 39 colleges making up Oxford University had had a significant impact on the world. Many scholars referenced the beginning of teaching in Oxford as dating back to the establishment of the first Abbey. Formal documented teaching at Oxford could be traced to 1096, making it the oldest English-speaking university in the world. It was the second-oldest university overall, behind the University of Bologna in Italy.

In the 13th century, the first University Chancellor had been appointed, and the first college, University College—known simply as "Univ"—had been founded. However, this detail was disputed by Balliol and Merton Colleges, as all three were founded during the 13th century.

The colleges resembled what one might picture in the mind's eye: antiquarian residence halls and meeting chambers, where students in long robes sat at long tables. The hierarchy of the college placed the Dons and Deans at a head table on a raised dais. Students were seated in rows at long tables running perpendicular to the head table so that all could see the power positioned at the dais.

The gothic buildings housing education, literature, research, administration, and worship appeared as they had for almost a millennium. The famous Bridge of Sighs, modeled after Venice's Rialto Bridge, spanned a cobbled street connecting two university buildings. Gargoyles smiled down upon students as they meandered through the lush, verdant English midlands and the Cotswolds.

This was the world Bella entered during Week 0 (the week before the term began), referred to as Freshers' Week, when new students descended upon the hallowed halls of the University of Oxford. Freshers' Week was filled with tours, orientations, and registration, culminating in the annual Freshers' Fair. This event allowed incoming students to explore the over 400 clubs and organizations suited to their interests outside their academic studies: sports, politics, literature, social events, punting, and, to Bella's excitement, a physics club.

The Oxford Physics Club was meeting in a conference room at the Bodleian Library, the principal—although not the oldest—of Oxford's many libraries. After signing up for the club, Bella crossed the quadrangle on her way to its first meeting. It was a rare, beautiful late summer's day, sunny and temperate, with just a hint of fall's cooling. This allowed her to wear her favorite red Aran cable knit jumper, skinny blue jeans, and matching red Repetto low-heeled ballet flats. Normally, the outfit was classic and proper, but on Bella, it resembled something out of a runway show as she crossed the quad. The eyes of both male and female Oxonians tracked her as she progressed.

She entered the library and took in its intoxicating atmosphere, which sunk deeply into her with a sense of joyful anticipation of the learning she was about to dive into. A gentleman holding the door for her carried a stiff, uncomfortable posture. She asked him for directions, and he stuttered, pointing a finger in the right direction.

"Thanks!" she said cheerfully before moving on, her energy perhaps a bit much for him.

She climbed the stairs, hearing the noise of conversations before she reached the meeting room. As she entered, she noticed the décor: polished wood, a large central table, and surrounding leather chairs that appeared both welcoming and intimidating to a Fresher. To Bella, it seemed like a place where Shakespeare and his contemporaries might have gathered.

The door squeaked as it closed behind her, and several heads turned in her direction. While a few continued their conversations, most of the noise stopped. The echoes of the quieter room seemed sharper. Bella calmly approached the group.

"Please," she said politely, raising her arms slightly. "Don't let me interrupt. I'm only here for the audition."

The previously talkative young men stopped speaking, their collective faces momentarily intimidating—even for someone with Bella's confidence.

An upperclassman eventually piped up. "Well, although this is an exclusive club, you've passed the audition!" He smiled warmly. "I'm Payman. Third year. Specializing in nuclear physics. Don't be put off. Join us. We're a friendly bunch."

"Not the geeks so often generalized?" Bella asked with a teasing smile.

Payman chuckled confidently, and the other men began stepping up to introduce themselves. However, Bella turned to the two women in the room first and introduced herself to them.

"I always appreciate women in science," she said warmly. "What are your names?"

A petite blonde with large glasses, responded comfortably. "Jenni, fourth year," she said. "But with my interests, I'll probably be here for a lifetime. Quantum physics and astrophysics—looking into new interstellar propulsion possibilities. Don't mind the guys. They're our mates, but women aren't exactly their forte."

The other woman, added with her right hand extended, "I'm Sarah, we women stand together, but honestly, you were right the first time. We're all a bunch of geeks; the guys just think only men can be intellectually gifted. They're playing above their league with us."

"Mean," Payman said casually. Bella smiled as Sarah pinched her tongue between her teeth.

"Really," Sarah continued, "we're all equals in this club. Sometimes women just have to take charge—and it's easy with these guys!"

"Okay," Jenni interjected, clearly amused. "Let's be nice and welcome our prospective members, even those who have 'already passed the audition.' If you're interested in physics, then you're in."

The ice was broken, and the budding physicists made their introductions. Jenni, who appeared to be something of a leader, discussed the club's objectives. It quickly became clear to Bella that this was more of an intellectual think tank than a pseudo-social club. The discussions went beyond standard physics topics like energy, force, and power. They explored the expanding universe, the bending of the space-time continuum, antimatter, baryons, charmed quarks, and quantum indeterminacy. Bella felt at home and welcomed in this rarified intellectual company.

She turned to Jenni with a smile. "All kidding aside, my reference to John Lennon's quote about the audition was more out of nervousness than anything. I wasn't sure if I'd fit in, but you've been so welcoming. The ideas are challenging, and if you'll have me, I'd like to join. I think I've found a place where I belong."

"Of course, you passed the audition," Jenni replied with a grin. "I look forward to getting to know you better, exchanging ideas, and challenging time and space. We girls need people like you. We stick together."

As Jenni looked around, it was clear everyone was excited and engaged. There were even a few laughs. Then she suggested they retreat to The Turf, one of the oldest pubs in Oxford, to continue their conversations.

CHAPTER 5
THE TURF TAVERN

The Turf Tavern Oxford, England
04 August 20:12 Local Time

The group laughed and chattered just like any other group of Oxonians. The men really were what might have been considered typical geeks—socially uncomfortable, boastful, and prone to challenging one another. The women, on the other hand, were comfortable with who they were and easily engaged in friendly conversation. Some of the guys displayed the typical immaturity of trying to attract the women's attention by interrupting, acting juvenile, or walking between them and laughing. One of them, Bella noticed, was different: Payman, the confident, swarthy third-year who had originally welcomed her.

As they walked together on Catte Street, Bella felt at home with the group. The guys reminded her of those she had grown up with in Spain and met while in school, but the women were different—they were far more self-assured and confident in their abilities, easily holding their own in complex dialogues. Their confidence made her feel welcome.

As the group turned onto New College Lane, Bella was awed as always, by the Bridge of Sighs, the skyway across the lane that symbolically and physically connected the old and new quads of Hertford College. Bella remembered the Rialto, the bridge of which this was a replica, from her visits to Venice. Seeing this version in Oxford always took her breath away.

"So, you like our bridge?" Jenni asked, bringing Bella out of her trance. "Legend says that lovers will be gifted with everlasting love if they kiss under the bridge at sunset. People say that legend came from the famous Rialto Bridge, but we like to think of it as our own."

Bella smiled. "I'll look forward to meeting someone worthy of everlasting love."

"There are attractive and intelligent guys in Oxford, just not here," Sarah quipped. "Here we are, just down this narrow alley. Hell's Passage."

Bella laughed quietly. "Sounds nice."

"It's actually St. Helen's Passage," Jenni answered. "It used to be Hell's Passage because of its narrow, somewhat hidden nature and its storied past, reflecting the difficult living of former times. But the name was changed to the more respectable St. Helen's."

Then at the end of the short passage, they arrived at The Turf Tavern, one of Oxford's oldest pubs. Running along one side of the tavern was the remains of the original city wall. The Turf had sprung up outside the city walls to escape the jurisdiction of the city and college leadership but had since been absorbed by the growth of Oxford.

Quantum

The Turf was crowded with new and returning students, townies, tourists, and tutors. Bella noticed Payman smiling at her across the table where the students drank ales and laughed at typical young adult jokes. Payman's blue eyes were mesmerizing to Bella.

Payman moved closer to Bella and launched into a history of the Turf. "See the plaque, this is the location where Bill Clinton made the infamous 'smoked but didn't inhale' while he was a Rhodes Scholar. There's the sign right over there with the exact location."

Bella felt a tingle she hadn't experienced since she was much younger and had met a young American on the Camino. Trying to keep herself composed, she listened to Payman, smiled, and looked directly into his blue eyes.

It must not have worked; Payman's brow twitched, and the skin around his eyes softened. "Are you still with me? Am I boring you with this little bit of the Turf's history?"

"No, no! Please continue," Bella insisted.

"Your pint is empty. Let me get you another."

"No, wait. I am absolutely fascinated. I don't normally speak so frankly, but I want you to know I am not bored, and I don't need another drink. How about you tell me something about you, not this building."

"Well, you're right; I never expected that. Let's move out of the aisle, and I'll share my story." Bella agreed, and they moved out of the aisle to a table for two that had just been vacated.

"I'm Iranian. My family has a long history of international engagement, and I was sent off to public school here in the UK. I loved it, I attended Oundle School. The public-school education paid off; I got into Balliol College and have found my passion in my studies. I know I'm lumped in with the geeks—and I love that. But my passions are traveling, skiing, and sailing."

"So then, why are you here?" Bella asked.

"To make my mark on the world, of course. I'm a third-year student interested in exploring alternatives to fossil fuels to meet the demands of the world and to try to save our planet since our elders just don't seem to give a shit."

Bella couldn't tell how much of his tone was ironic but stopped thinking about it when she caught his smile. "Ah," was all she said.

"What about you?" Payman then asked. "Who are you, Bella?"

Bella found herself softly biting her lower lip. She felt weak in her knees and took a deep breath. "I have a dream to complete my DPhil in quantum physics and understand the most basic building blocks of the universe. I applied to the University and, luckily, was accepted to Balliol because it is the best of Oxford's colleges for physics and research. Both of us are special as members of this elite college. As you know, there are over 80 applicants for each place."

"Are you thinking of measuring up to the greatness of Oxford alumni past?"

Bella nodded. "Yes, that's my dream. Sir James Jeans and one of the pioneers of quantum physics, Max Born, are two in particular."

She took another deep breath and calmed herself once more. She felt she may have gone too far with the last statement. How could she ever measure up to Max Born? The low hum of bar chatter helped to soothe her.

"We have a lot to live up to."

"We really do; the pressure is always there," Payman agreed. "My family has very high expectations, and I get regular visits from my grandfather, who has offices in London. He reminds me of what he holds as my future—becoming famous as a scientist while running the family financial businesses. The two are polar opposites, but not to him. It's all about achievement."

Bella gestured for him to continue, partly hoping that it would keep his focus on her and not on the antics of the others in their group.

"My family ancestry is actually very old. We can trace it back many generations to the Persian merchants who expanded trade around the eastern world. We were sailors who traveled the seas to trade fruits, nuts, and silks across the Indian Ocean, the Mediterranean, and the many seas in between. My grandfather is in finance and spends time in both London and Plymouth when he is in the UK. He taught me to sail, and we have an amazing ocean-going sailing yacht harbored in Plymouth. I crew for him when we race."

"You race?" Bella suddenly wished she'd accepted Payman's offer for a drink—something to occupy her hands.

"Recreationally. Someday, I'd like to competitively sail in major international regattas." Payman paused. "Sorry, I'm probably boring you, but you asked."

Bella smiled. "Nothing boring about a man of depth and influence who is also very smart."

The two continued their conversation, sharing stories of their youth and upbringing. Bella shared her love for the Camino and urged Payman to experience it and visit her mom's café. They learned they both had a passion for skiing. Bella was surprised to learn that Payman had learned to ski in Iran but had skied many of her favorite ski areas across the Alps. Both seemed infatuated with one another. Bella thought, Is there really such a thing as love at first sight? The scientist in her countered, It's more about chemistry.

The spark grew into a flame. They lost all track of time, not noticing that almost all the others had left. They hadn't said their goodbyes and became self-conscious, focused only on one another.

"Looks like we were so focused on personal discussions that we missed the technical dialogue of our colleagues," Payman said with a smile. "But I'm not sorry for that. I've really enjoyed getting to know you. You seem so much more experienced and have lived beyond your years. I must say I'm really happy we met. I want to see more of you."

"Me too," Bella replied. "You can see me home. That would be appreciated, and it would be in keeping with your desire!"

CHAPTER 6
BALLIOL COLLEGE

Oxford University Oxford, England

September 3 years later

Three years passed quickly. Bella and Payman had become an inseparable pair, both at Balliol College and in the elite world of physics research. Since Bella's first day on campus, their relationship had grown from a professional friendship to a romantic partnership. It seemed destined to lead to groundbreaking publications, prestigious positions, marriage, children, and a perfect life at Bella's family estate in the Cotswolds. At least, that was Bella's dream. Payman always listened to her hopes and encouraged her imagination.

Bella quickly proved herself as a brilliant researcher. She aligned with one of the world's leading quantum physics experts, earning her place in this challenging field. For most people, quantum physics was a confusing subject that made little sense. It asked people to rethink everything they knew about the universe, from tiny particles to massive galaxies. It included strange ideas, like a cat being both dead and alive at the same time or particles being waves and vice versa. But Bella and Payman not only understood these concepts—they thrived in them.

Quantum physics focused on the smallest building blocks of matter and energy, like atoms and subatomic particles. These went beyond familiar particles like protons, neutrons, and electrons to include things like quarks, antiparticles, and the mysterious Higgs boson. Predicted in 1964 but only confirmed in 2012, the Higgs boson—dubbed "the God particle"—was essential because it gave particles their mass. If matter and antimatter met, they would destroy each other in a burst of energy. Thankfully, antimatter was so rare it wasn't a real-world problem.

In Angels & Demons, Dan Brown catapulted antimatter into the global imagination. What had once been the domain of particle physicists became a household mystery—a substance so volatile that a single drop could flatten a city. Brown's story blurred fact and fiction, suggesting that antimatter, produced beneath Geneva at CERN, could be weaponized and stolen by a secretive brotherhood.

The truth? Antimatter is real. It's the mirror opposite of matter, and when the two meet, they annihilate in a burst of pure energy. But it's incredibly rare, produced in microscopic quantities at great cost. Still, the idea stuck. Antimatter became a symbol—a modern alchemist's dream, a sci-fi doomsday device cloaked in scientific credibility.

And at its center was the Higgs boson, the elusive "God particle" that gives mass to both matter and its darker twin. It was proof that the universe has structure—and perhaps, secrets better left undisturbed.

The behavior of these tiny particles challenged traditional science. Early physicists couldn't believe things could exist in two states at once or be in two places at the same time. But experiments like the double-slit experiment proved it: particles acted like both waves and particles, depending on how they were observed.

One key idea in quantum physics was entanglement, it described how particles could become linked so that changing one instantly affected the other, no matter how far apart they were. This idea baffled even Albert Einstein, who called it "spooky action at a distance." It wasn't until years later that scientists confirmed entanglement was real.

Quantum mechanics also led to exciting developments like quantum computers. These computers used qubits, which could exist in multiple states at once, allowing them to solve problems far faster than regular computers. By using entanglement, quantum computers could perform incredibly complex calculations.

For Bella and Payman, the strange world of quantum physics often mirrored their relationship. They worked on similar ideas but in different areas. Bella focused on subatomic particles and collaborated with Oxford's team and CERN, while Payman explored peaceful uses of nuclear energy. Despite their different paths, they remained deeply connected, almost like entangled particles themselves.

Their connection often felt like Einstein's "spooky action at a distance." Bella found her relationship with Payman both exhilarating and overwhelming. When they were together, she was happiest, but when apart, she missed him intensely. Yet, in those moments, it always seemed as though Payman somehow sensed her longing, as if they were in sync across any distance. For Bella, the memory of being with Payman was pure bliss.

CHAPTER 7
PICNIC

Oxford Botanic Gardens *Oxford, England*
23 October 14:20 Local Time

In their years together, Bella had struggled to get Payman to leave the confines of the labs, pubs, libraries, and her home. On a warm, beautiful fall day, she finally succeeded. Payman agreed to her request, and she managed to pull him away from his lab.

One of her favorite places was the Oxford Botanic Garden, the oldest botanical garden in the United Kingdom. Founded in 1621, the gardens had long provided a place for lovers to explore nature and deepen their relationships, walking hand in hand through the Glass Houses, flower beds, gardens, and miles of pathways filled with some of the most diverse tree specimens in the UK. Bella spotted one of the many pop-up shops offering handmade treasures to visitors, and something particular caught her eye.

Look at these beautiful baskets—so perfect for a picnic," she said, barely containing her glee. "I'm going to get one." Bella and Payman then explored their first purchase together—a lasting investment in their future: a handwoven, one-of-a-kind picnic basket.

"Bella, how do you know we'll have time for picnics?" Payman asked.

"Well, if you won't picnic with me, I'm sure one of the physics nerds would love to discuss Higgs bosons over a nice Sauvignon Blanc."

"You really know how to make a guy jealous. I'm not letting anyone else whisper sweet nothings about bosons on a picnic blanket with my girl."

Bella's cheeks flushed; it was the first time he had referred to her as "his girl." "In that case, you'll have to help me pick the one we'll use together."

She examined the baskets, fascinated by their quality. The wicker was delicately but tightly woven into bright, braided patterns, and she couldn't take her eyes away from one in particular. It was a traditional set that included everything needed for a Victorian-style picnic, even an oversized blanket. Smiling to herself, Bella made the purchase, imagining the adventures and romantic moments she would share with Payman and her new treasure.

They carried the basket and continued exploring the gardens. Bella wasn't done persuading Payman to indulge in her idea of fun. She took his hand and led him along less crowded paths until they found a secluded grassy patch under a canopy of pines.

"This is the perfect spot for our first picnic," she declared.

"But we didn't bring any food," Payman replied. "No problem. We need to test the blanket anyway. Well, let me help you with that."

Bella couldn't resist diving into his earlier comment about her being "his girl." As much as she loved Payman and as easily as they communicated in physics, in the language of love, Payman, in Bella's words, was a little clueless. She was curious to see how vulnerability and affection looked in his world. They spread the blanket and sat down.

"This really is a nice spot," Payman said.

"Thanks." Bella stretched her arms and back, then slouched comfortably into the blanket. They both laughed.

"Vain," Payman teased. As they continued talking, Bella felt her chest tighten with anticipation. She wanted to know more about him, to explore the depths of their connection. Still, their conversation naturally drifted back to science, their work, and observations about the English countryside. The forest carried a musky, natural aroma that seemed to awaken something erotic in her. Old branches hung above them, creating a canopy of dappled sunlight. Eventually, Bella thought, If I want this to happen, I'll have to take the lead.

She reached for Payman's hand and pulled him closer. "It's time we test the comfort of this blanket." Bella's desire pushed aside her thoughts of small talk, and she gave him a gentle kiss. Relief and longing washed over her in equal measure.

Payman took a deep breath, and she felt his chest soften. A shiver ran down her spine, signaling that something was about to happen.

He kissed her back, his lips warm and inviting. Bella quivered as their bodies came closer, their kisses deepening with passion. The dappled sunlight around them seemed to dim as they focused entirely on one another.

She pulled him down to lay on the blanket and wanted to show him how much he meant to her. She was hesitant, there were families within hearing distance but she couldn't resist. She wrapped one leg around him and moved her hips to make sure he was aroused. She felt him, he was already hard. She wore nothing under her skirt. She wanted him inside her and he knew it.

A moan emerged before she could stifle it. Payman gently kissed her and whispered, "enjoy but stay quiet. We don't want to be interrupted by a couple pushing a pram." She nodded with her eyes closed and pulled her lower lip between her teeth to remind her to enjoy in silence. Bella wrapped her legs around him. He was restrained by his pants barely below his thighs. But it worked. So he succumbed to her overwhelming insistence and allowed her to lead their progression. She put her hand over his mouth to ensure a reminder of silence. Losing each other in the other, all time slipped away, they were the only two on the planet. She took charge. In their world of each other, she brought him to climax. She herself enjoyed several orgasms.

The two collapsed into one another as one being. She held him as she wanted the two to remain one being forever.

As they continued to embrace, the outside world faded away. Their connection, both emotional and physical, felt complete. They lay together, lost in the moment, exploring their lasting intimacy.

The sound of guests nearby brought them back from the brink of erotic recovery. They smiled, chuckled, and sat back up. No words could express their sentiment. They just sat in silence. Holding one another and appreciating the moment.

Hidden in the brush, they composed themselves. Bella brushed her skirt into place and reached a hand to Payman. "OK champ, time to get up off the mat. I've completely conquered you." He pulled his hand away as he struggled to get his jeans back in order.

Payman laughed, struggling to pull his jeans back into place. "Easy for you. All you have to do is stand. I'm struggling here." Together they smiled and laughed, and as he stood up, they kissed again. They busily put their nest back into the picnic basket, made it out of the brush, and joined the others strolling through the park. With a smile, they shared their secret, reached for one another's hand, and slowly meandered while enjoying the day.

CHAPTER 8
PAYMAN'S ABSENCE

Around Oxford University Late October

The week following their first amorous picnic, Bella could only look forward to having Payman alone again. Her education and her classes gave her no rest from her internal challenges and demands. On the contrary, her desire for Payman and her love of learning were in lockstep for her attention, and focusing on her classes took much more energy and focus than she'd ever had to exercise. She put all her abilities and energy into achieving the demands of a graduate student at Oxford.

As time passed, Bella's classes and research became all the more demanding. As a lab assistant, she supported the research of one of the world's leading subatomic scientists, who was also a mentor. Research kept her in the lab most nights—she and Payman had little time to see one another. Her desire for Payman was put on a back burner and would just have to wait.

Then surprisingly, Payman was called back to Iran to attend to a family issue. His research could be transitioned to a laboratory in Iran and his graduate work would continue. He was making significant advances and both his sponsors in Iran and his dissertation committee supported his continued work while away, so his absence could reasonably be indefinite. This was a punch to her gut. She would miss him and showed him as much, but as the overachiever, she knew she needed time to focus on her own studies and research. This time apart was enough rationale for her to properly focus.

With Payman away, Bella fell into her other passion, work. She visited the many libraries of Oxford to find a quiet place to study. She explored study centers and library carrels but found she was drawn to the warmth and history of the Bodleian Old Library. Her favorite was a quiet corner of the famous old library, Duke Humfrey's Library with its silent study conditions and its comfortable yet practical setting. It was originally constructed in the late 15th century and was named after the man whose donation of around 281 manuscripts laid the foundation for the library's collection.

Bella loved Duke Humfrey's for its beautiful Gothic architecture; the intricate woodwork, vaulted ceilings, and tall, narrow windows that allowed natural light to filter in. The interior retained much of its medieval charm, with dark oak bookcases, reading desks, and a sense of historical continuity. It featured a series of ceiling paintings and carvings that depicted the coats of arms of the university and its benefactors. The reading room provided an atmosphere for quiet study, surrounded by centuries of scholarly tradition. The smells were a complex and evocative mix of musty old books, rich wood, leather, and library dust, creating an atmosphere that Bella could fully immerse herself in and allow her to sink into both history and knowledge.

She became a fixture in Duke Humfrey's when not in the lab. She would work until the work became too intense and she needed a people fix. She would go to her favorite place for a coffee and a more social read at Blackwell's Books across Broad Street from the Bodleian. Savoring coffee and a favorite novel. Giving her the ability to recharge her batteries.

The physics of attraction didn't end when Bella recharged her batteries. She noticed the handsome bartender from the Kings Arms. His natural magnetism challenged her. He smiled as he approached, "I know you from the KA. What are you doing here? This is my study spot. But I'll share if you will?"

Bella sensed the come-on, but the magnetism was too strong for her to resist. She jumped right in. "I like the concept of sharing. I've always played well with others." Her cheeks flush and she feels an attraction."My name is John. I'm reading chemistry. Can I sit at your table and finish my assignment? But most important, after we finish our work would you join me for dinner and pints?"

Bella, normally the assertive one, liked John's approach. "Since Duke Humfrey's is your favorite study spot, mine too. You're a scientist and I love the KA, we must have even more in common, so yes. Let's get our work done and then explore sharing opportunities."

The two attempted to focus on their work in the library, but they found the flame of their Bunsen burner just had grown too hot. Each caught the other peeking or even staring with a "come hither smile". Finally,. The anticipation was beyond what John could withstand.

"I'm finished if you are?" Bella just smiled and said yes and connected to John with her smile and her eyes. It was natural magnetism. The two young university students left the library, they appeared no different than any other young couple heading to the KA for an evening of frivolity and maybe more!

PART 3

CHAPTER 9
THE VICTORY

Yacht Club Costa Smeralda's (YCCS) Porto Cervo, Sardinia, Italy

09 September, two years later 17:30 Local Time

The 'Falkin and its crew celebrated their victory as they cruised into their slip. Their triumph in the Maxi class of the Rolex Cup was the pinnacle of yacht racing and the ultimate aspiration of every competitive sailor who yearned to reach the top of the sport. The crew basked in the knowledge that on the water, the level of competition had been of the highest order. The race courses were some of the most challenging and spectacular in sailing, offering an unrivaled combination of competition, conditions, and scenery. Held at Yacht Club Costa Smeralda, off the coast of Porto Cervo, Sardinia, it was one of the most stunning and diverse places in Italy. Nestled in the Mediterranean to the west of the Italian mainland, the five days of competition had offered a program of primarily coastal races, allowing the fleet to navigate some of the world's most impressive natural courses. The action unfolded in alluring waters, set against the wild, rocky islands and islets of the Maddalena archipelago.

The Maxi Yacht Rolex Cup embodied the very essence of superlative yacht racing. The environment, the organization, the yachts, and the participants were all exceptional. The crews comprised the world's foremost sailors, a veritable who's who of the yachting world. Even for those accustomed to success, victory at the Maxi Yacht Rolex Cup remained a significant achievement.

On the 'Falkin, the crew was a mix of the owner's friends and professional racers, led by British and French sailors experienced in the America's Cup and around-the-world races. In the true Corinthian spirit, the team celebrated their victories with a wide variety of spirits. When in Italy, Asti Spumante was the victory celebratory beverage, and the crew had happily partaken. Spirits weren't the only draw of the evening; the natural beauty of the local women—many of whom were world-class sailors in their own right and present at the club for the post-race excitement did not go unnoticed.

Lewis was Payman's best mate on land and his first mate on the 'Falkin. The two had known each other since their days at Oundle School, a primary school in the quaint historical town of Oundle, England. The two were opposites in almost every way, and when this was pointed out, one, or often both of them would simply respond, "Opposites attract."

Lewis was all muscle yet lanky, with fiery red hair and emerald-green eyes. An extrovert by nature, he was loved by everyone and had a natural magnetism. The life of the party, Lewis was not equally blessed with intellectual gifts. He attended Oundle largely because of his family's influence rather than his academic abilities and was sometimes referred to as 'frivolous.' Despite their differences, Lewis had been by Payman's side since they met.

Lewis had desperately wanted to be a sailor and asked his father, a wealthy import/export businessman, to allow him to spend summers learning to sail with Payman. Reluctantly, his father had agreed, with the condition that Lewis maintain a 5 rating in all his classes. Apparently, that was all the incentive Lewis needed. Though it had been a challenge, with Payman's help, Lewis had worked hard and achieved the required 5's. Payman, who regularly received 8's and 9's in his ratings, had happily tutored his friend through the process.

Once on land, the crew of the 'Falkin celebrated under the warm but waning glow of the Mediterranean sun. Porto Cervo, nestled along Sardinia's rugged coastline, was alive with energy. The hum of luxury yachts echoed across the marina, their polished hulls gleaming in the light. The salty tang of the sea mingled with the rich aroma of freshly grilled seafood and the faint sweetness of blooming jasmine lining the streets. Nearby, glasses clinked as other teams toasted their efforts, laughter and lively conversation filling the air like a melody.

Beyond the bustling marina, the Maddalena archipelago stood as a striking backdrop, its rocky islands and jagged cliffs painted in hues of gold and orange as the sun began its slow descent. The evening light cast dramatic shadows over the wild landscape, creating a breathtaking contrast against the azure sea. Small fishing boats bobbed gently in the calm water, their nets stretched out to dry, while seagulls circled above, their cries blending seamlessly with the sounds of celebration.

At the Yacht Club Costa Smeralda, the energy was palpable. The iconic building, with its sleek white façade and terraced balconies, offered sweeping views of the harbor. Inside, the club buzzed with activity as sailors, locals, and visitors mingled. The air was a tantalizing mix of salt, perfume, and the earthy undertones of Sardinian wines. Candlelit tables spilled onto the patio, where couples leaned into one another, their conversations intimate yet animated.

As Payman took it all in, he couldn't help but feel a sense of awe. The location was unlike anywhere else—exceptional in its natural beauty, yet welcoming in its spirit of camaraderie. The victory combined with the setting made the moment unforgettable. Above, the sky shifted into a gradient of deepening blue, stars beginning to shimmer faintly against the backdrop of the twilight.

Lewis greeted Payman with a bro-hug and a bottle of Spumante. "Cheers, mate! We did it! Oh my God, we did it!" Lewis beamed with pleasure and pride.

"We did it together," Payman replied. "It's really our celebration."

Lewis turned to the gathering crew and raised his glass. "To our wives and sweethearts!"

The crew responded in unison, "May they never meet!" Laughter erupted as glasses clinked. Payman, loyal to Bella, blushed at the toast but chose to highlight the crew's accomplishments.

"As with any contest of speed, precision is key," he said. "We raced with precision and determination. With several Super Maxis withdrawing due to technical failures, this was a war of attrition. A big hand to the crew and Chris, our maintenance director, for ensuring everything stayed in perfect working order."

Laughter and cheers echoed through the crowd as Payman raised his glass again. "To the best sailing team on any sea!"

The crew erupted in cheers, their voices blending with the lively marina ambiance. Payman continued, "Coming to Sardinia, working together as one, we conquered one of the most challenging courses we'll ever face. Now let's enjoy this Sardinian evening!"

"Here, here!" the crew roared, their camaraderie unmistakable.

Payman chuckled as he spotted Lewis returning, flanked by two stunning local women. Lewis, ever the charmer, introduced them. "Payman, meet Francesca and Maria. Ladies, this is our skipper and my best mate, Payman. They've been cheering for us every day!"

The women laughed warmly. "You both enjoy sailing then?" Payman asked.

"Hell yeah!" Maria replied enthusiastically, her hazel eyes sparkling. "We both sail on our own team. The races here were incredible."

Payman admired their athletic builds, noting the strength in their shoulders and arms, certainly sailors. "And now," Lewis interjected, "we're off to Francesca's family restaurant for the best seafood you'll ever taste, made with love by her brothers, the chefs!"

Payman, caught between the emotions of victory and the charm of the two women, smiled. "Hello, Francesca and Maria. It's a pleasure to meet you both. I thought our ship was the most beautiful thing in the world, but you've made me rethink that."

The women laughed, and even Lewis looked impressed. Payman continued, "How about joining us on board for a toast?

Lewis can give you a tour of our victorious but slightly messy boat."

Lewis eagerly agreed, saying, "Can't turn down an invitation from the captain, surely."

As they boarded the 'Falkin, the women showed their appreciation for what they saw, while Lewis enthusiastically demonstrated its features. Moving from the racing deck to the creature comforts, they explored every corner. There were two identical berths in the bow, one each for the captain and the first mate. Lewis, eager to impress Maria, quickly showed her his bunk. To Payman's surprise, it was perfectly made and spotless—Lewis must've enlisted a deckhand and had tidied it beforehand. Now Payman understood; Lewis clearly had intentions beyond the usual camaraderie.

All around the deck, the crew bustled with energy. Some were cleaning and stowing gear, while others prepared the boat for the journey home to Plymouth. Several of the crew had been sent ashore to shop for the freshest vegetables, meats, and, naturally, a generous supply of Italian wines.

Amid the flurry of activity, Chris, the maintenance director, approached Payman with a rare grin. "Skipper, I think we deserve some kind of engineering medal. Keeping this beauty in one piece during those winds? Miraculous."

Payman smirked, clapping Chris on the shoulder. "You pulled it off, Chris. A well-oiled machine, thanks to you."

Nearby, one of the younger deckhands, Marco, leaned against the railing, catching his breath. "Captain," he called out with a cheeky grin, "thanks for not pushing us too hard out there. Easiest race of my life."

The crew erupted in laughter, fully aware that they'd worked harder than ever before. Another sailor, Jake, chimed in, "Yeah, Marco here only missed two trims this time."

"Three, actually," Marco admitted, raising his hands in mock surrender.

Payman chuckled, shaking his head. "You all gave it everything. That's why we're standing here, victorious. Well done, everyone."

The camaraderie was touching, the kind of bond forged only through shared trials and triumphs.

Later, to allow the crew some downtime and space to prepare the boat, Payman, Lewis, and their guests stepped ashore. The streets of Porto Cervo beckoned them with their winding cobblestones and glowing lamplights, casting a soft, golden hue on the picturesque town. The warm air carried the mingling scents of saltwater, freshly baked bread, and blooming jasmine. Musicians played lively tunes at corners, their melodies weaving through the laughter and chatter of locals and visitors alike. Some free spirits danced as the music moved them.

As they strolled, Francesca took the lead, guiding them through narrow alleys that opened to breathtaking views of the harbor. The group paused momentarily at a quaint square, where a fountain trickled serenely. Children chased one another in circles, their giggles adding to the charm of the scene. Payman found himself admiring the quiet elegance of the town, its simplicity contrasting beautifully with the adrenaline-filled race earlier in the day.

"This way," Francesca said, turning down another narrow street. She gestured to a warmly lit building with colorful awnings. "Welcome to my family's restaurant."

The small establishment exuded warmth and authenticity. A wooden sign above the door read Trattoria della Costa, and the air was rich with the aroma of garlic, olive oil, and fresh herbs. Inside, the atmosphere was intimate, with rustic wooden tables adorned with checkered cloths and small vases of fresh wildflowers. The walls were lined with photographs of the Maddalena archipelago and Francesca's family, capturing decades of love and hard work.

As they entered, Francesca's parents, a jovial couple with kind eyes and welcoming smiles, approached with open arms. "Ciao, benvenuti e congratulazioni ai campioni Rolex! Benvenuti!" her father exclaimed, shaking Payman and Lewis's hands firmly. Her mother embraced Francesca warmly before turning to Maria with equal enthusiasm.

"Francesca has told us about your great victory," her mother said in accented English. "Tonight, you will eat like champions!"

Even as an outsider, Payman could sense the pride and hospitality radiating from Francesca's parents, who treated them like VIPs. They were ushered to a corner table near the open kitchen, where the aroma of sizzling garlic and the clatter of pans added to the lively ambiance. The restaurant buzzed with energy as diners raised glasses and shared hearty laughter, a celebration of life's simple pleasures.

The meal began with a selection of antipasti: marinated artichokes, sun-dried tomatoes, olives, and freshly baked focaccia drizzled with olive oil. The group toasted with glasses of crisp Vermentino, the local white wine, and the conversation flowed as easily as the wine. Francesca's brothers, the chefs, occasionally peeked from the kitchen to wave at the table, their faces beaming with pride.

When the main course arrived, it was nothing short of a masterpiece. Payman's seafood and vegetable risotto glistened with a medley of colors: vibrant orange shrimp, tender white scallops, and the deep green of freshly chopped parsley. The dish was paired with an unusual yet exquisite red wine, its rich, earthy notes complementing the delicate flavors of the seafood. Payman took a bite and closed his eyes, savoring the harmony of textures and flavors.

"This is divine," he said, setting his fork down momentarily. "I'm beginning to think your chef might actually be a minor deity."

Francesca laughed, her cheeks flushing with pride. "That would be my brothers. They'll be pleased to hear that. It will swell their heads."

As the evening progressed, the group felt increasingly at home. Francesca's father visited the table with tales of Sardinia's history and the Maddalena archipelago, his deep, sonorous voice drawing them in. Maria and Lewis engaged in playful banter, their laughter frequently drawing glances—and smiles—from nearby tables. Payman, ever the introverted gentleman, listened more than he spoke but found himself charmed by the warmth and simplicity of the evening.

When dessert arrived—a platter of seadas, traditional Sardinian pastries filled with cheese and drizzled with honey—the group couldn't resist digging in, their conversation pausing briefly as they indulged in the sweet, tangy treat. Francesca assured them the meal was on the house, but Payman, ever gracious, discreetly left a handful of 100-Euro notes on the table for the servers and kitchen staff.

As they exited the restaurant, their spirits were high. A group of locals, clearly impressed by the 'Falkin's victory, raised their glasses in a toast. "Salute! Buonanotte ai vincitori della Rolex Cup! Meraviglioso."

Lewis, ever the charmer, took a proffered glass of spumante, clinked glasses with the men, and, to everyone's surprise, replied in flawless Italian, "Grazie. Saluti. Ora aspettiamo l'anno prossimo! Arrivederci per ora. Ciao." (Thank you. Cheers. Now we look forward to next year! Goodbye for now.)

On the cobblestoned street, the warm night air embraced them, and the faint sound of waves lapping against the harbor added a soothing rhythm to their walk. Payman glanced at Francesca, who walked quietly beside him, her smile soft and genuine. For a moment, he allowed himself to appreciate the simple beauty of the evening—the camaraderie, the celebration, and the joy of a shared victory in such an enchanting place.

The group strolled the cobbled streets of Porto Cervo under a sky thick with stars. The warmth of the day lingered in the air, and the distant sound of waves lapping against the marina added a soothing backdrop to their quiet laughter. Francesca slowed her steps as the harbor came into view, letting the others walk ahead.

"You know," she began, her tone light but probing, "you're not what I expected from a racing skipper. Aren't captains supposed to be loud and commanding, like something out of a pirate movie?"

Payman chuckled softly, shoving his hands into his pockets. "I've found that a calm captain usually has a calmer crew. Besides, I save the theatrics for the big wins."

She stopped and turned to face him, her hazel eyes searching his. "So, what's the act now? Serious captain, heading straight back to the ship?"

His smile faltered slightly, and he hesitated before answering. "A good captain knows when to celebrate and when to prepare for the next leg. And for me, that's tomorrow."

Francesca tilted her head, her lips curving into a faint smile. "You're a hard man to read, Payman. One moment, I feel like I'm getting to know you, and the next, you're back to being this perfectly polished skipper."

He met her gaze, his Mediterranean blue eyes softening. "Maybe I just like to keep people guessing."

She let out a light laugh, but there was a trace of disappointment in her expression. "Well, you've certainly succeeded. I don't know if I should feel flattered or frustrated."

Payman stepped closer, his voice dropping slightly. "You should feel neither. You've been a wonderful part of this evening, and I'm grateful for your company. But as captain, my responsibility is to my crew and my ship. They depend on me."

Francesca lingered for a moment, her gaze held his. She nodded, her playful demeanor giving way to something more sincere. "Well, for what it's worth, I think your crew is lucky to have you. And so is… whoever has your heart."

Payman blinked, momentarily caught off guard. He recovered quickly, a faint smile playing on his lips. "Thank you, Francesca. That means more than you know."

She leaned in and kissed his cheek, her lips warm against his skin. "Buona notte, Captain," she whispered. Then, without another word, she turned and walked back toward the restaurant, leaving him standing there, watching her until she disappeared into the soft glow of the streetlights.

For a moment, Payman remained where he was, his thoughts tugging him in conflicting directions. The echoes of Bella's laughter filled his mind, grounding him in the loyalty he held so dear. Lewis and Maria were nowhere to be seen, off on their own festivities. With a deep breath, he turned and made his way back to the 'Falkin, ready to face the morning and all it would bring.

CHAPTER 10
BACK ON THE 'FALKIN

Yacht Club Costa Smeralda's (YCCS) Porto Cervo, Sardinia, Italy

09 September 21:45 Local Time

That evening, back on the 'Falkin, the crew was still basking in the glory of the Rolex Cup victory, one that would be proudly displayed on their competitive resumes.

Celebratory drinks, meals, toasts, and parties now gave way to the necessity of preparations for the week-long cross of the Mediterranean, run up the North Atlantic along the coast of Portugal to Finisterre, and across the Bay of Biscay to Plymouth. As much as the crew enjoyed the competition, the victory, and the celebration, they also had the discipline to ensure the ship was ready for its return.

Maurice La Crux was a ruthless and experienced sail racing strategist, who worked seamlessly with Clint Patterson & John Hind, two seasoned and enthusiastic sailors. Payman loved them as crewmates as they always gave their everything, only their full measure of effort. They busily readied the ship, even ordering other crew members to tend to their specialties. Of course,

Maurice still attended to every detail of the boat, ensuring that everything was in place and all systems a "GO".

Clint fancied himself as a chef of sorts, making the best he could with what was available. In Porto Cervo, his job couldn't be easier with the abundance of excellent seafood, fruits, vegetables, and baked goods.

John was the self-appointed wine aficionado. According to him, the world of Italian wines could be daunting. In a country smaller than the state of California, it was said that there were over 2000 unique native grape varieties and about 400 of those were used in commercial winemaking. Every province in Italy produced and exported wine, so there was always an overwhelming number of regions in which to keep track. The top choice of the year was apparently a Livio Sassetti Brunello di Montalcino Pertimali from Tuscany, one of the most prolific wine regions in Europe. When it came to history and tradition, no Italian wine region could compare to Tuscany, where viticulture had been practiced since the 8th century BC.

The rainbow-colored rolling hills and dappled afternoon light had made Tuscany the setting for many romantic films and famous works of art. These also formed the spiritual home of Italy's signature grape, Sangiovese. The wine-producing region Brunello di Montalcino was where Sangiovese truly shone.

For white wine, John chose Benanti Etna Bianco, sort of like a Chablis of Italy but with a personality all its own. It was 100% carricante grape, unlike anything found outside of Italy. It was grown on the volcanic soils of Sicily's Mount Etna, which produced a rich texture and a vibrant, almost salty acidity, with crisp citrus and orchard fruit flavors. This wine was particularly popular with fresh seafood, of which there would be an abundance on the trip home. John decanted a bottle of Brunello for the trio to enjoy while they finished their duties. Just then, Payman returned to the ship, smiling from ear to ear.

John called out, "Hey skipper, that grin tells me your evening with the Italian beauty didn't disappoint."

"On the contrary, I'm just happy to be here back with the crew," Payman called back.

"Yeah, yeah." John sarcastically rolled his eyes.

"Of course, nothing wrong with her. She knows how to enjoy life and food, but alas, I am better here with you and the boat than in town with a beautiful woman. Plus, I told her about Bella."

"Skipper, you know when you tell a woman about another woman, that's a buzz-kill. I don't know why you continue to sink your chances of an Italian romance."

"She was a one-in-a-million," Maurice said, "but Bella is a one-in-ten million. I trust your judgment, but you have passed up a real celebration of our cup victory."

"Let's toast to our victory and not dwell on what may have gotten away. We won the cup today!" Clint poured himself and the others a drink and raised their glasses, his voice slurring slightly. "To the best crew on the seas!"

Unlike most yacht owners, Payman wanted to spend his time on the cruise on the water with his crew. He enjoyed embracing the life of a sailor and the company of friends. After congratulating the crew one last time and taking a final sip of his drink, he called it a night. Retiring to his tiny but luxurious berth for a well-deserved sleep, he reflected on the day's many successes, both small and large. How could life be anything but wonderful? With

Bella by his side and the wealth his family had created, he felt truly blessed. He said his prayers and thanked Allah for his many blessings.

On deck, the crew wound down too, their laughter now softened by the rhythmic lapping of the waves against the hull. They toasted one another and their captain again, glasses raised under the starlit sky, as they reflected on their many blessings. As the night deepened, the ship stood readied, its sails stowed and provisions secured. The crew, now relaxed and content, let their emotions simmer with the quiet satisfaction of their triumph. One by one, they retired, the echoes of celebration giving way to the gentle murmur of the sea. For each of them, sleep came easily, filled with dreams of victory, smooth seas, and the allure of beautiful Italian women.

CHAPTER 11
SMOOTH SAILING

Porto Cervo, Sardinia, Italy

10 September 08:10 Local Time

The morning dawned with another beautiful Mediterranean blue sky and Lewis running down the pier to the 'Falkin with a smile from ear to ear to meet his mates and set off home. As he approached the boat he was greeted by jeers and questions from his mates. Everyone wanted to know where he had been through the evening. The question was more of a courtesy, however. He was not one to tell stories of his conquests and failures with women but his afterglow beam was there for them all to see.

John was the first to provide a fist bump and ask the question. "Was it worth it? You bypassed an evening with your mates for a quickie, you little hound?"

"It was not a quickie," Lewis replied, his teeth showing behind his smile. "It was an intimate, delicate night."

Most of the crew laughed.

"However, apropos of nothing, I need some rest. I hope you're all up to getting us underway so that I can get some sleep."

Maurice smiled and said, "By the way you look and the way are walking, you might need some Dramamine too. The hangover is going to be far worse than any waves we have to deal with. You know you've got work to do on this return but, because I'm such as a good mate, I'll cover you."

Following the exchange, and with the outgoing tide, the crew pushed off for their return home, and they picked up the perfect breeze to carry them into the Mediterranean. Payman sat amongst his team, quietly enjoying company with part of his mind on another. He observed the calm sea and felt the steady breeze—about 30 degrees off-beam, about 12 knots velocity. This was the ideal wind and most efficient angle to get back to England. To Bella.

After several hours, the water was still smooth the breeze was fresh. The only sound was the craft slicing through the water. Payman held the course as he adjusted the sails and watched the telltales fall into place. He took a moment, however, to watch himself. This wasn't his first sail, his first time with his crew, or even his first victory, but something about the past day made him wish he could etch this memory onto something as one for his lifetime highlight reel.

Maurice joined him in the cockpit, breaking the spell of contentment that had enveloped Payman.

"Million dollars for your thought!" piped Maurice.

Payman replied, "The only million dollars you'd get is for that cup of coffee you thoughtfully brought for me."

"You needed it then? Despite your relatively cordial demeanor the past day? If I didn't know any better, I'd say our captain is reaching a milestone in maturity."

Payman snorted. "It'll be a while yet. We're all still young." They watched the waves together for a while, Payman returning to his previous constellatory state, before he asked Maurice, "Have I ever told you the story of this watch?"

Maurice turned to him slowly. "No."

Payman glanced at his wrist, the glint of his grandfather's Rolex Yacht Master 2 Regatta chronograph,

catching the sunlight – a piece of history tied to his seafaring roots.

"This was a gift of celebration upon my first major regatta victory, given to me by my grandfather. It was originally his. He took it off his wrist and said it was now mine. He and generations of our family were seafarers. Grandfather told me often how they used to feel more at home on the sea than on the land, and it was he who taught me to sail. He was probably the most influential family member in my life. You could argue he made me who I am.

"I have tried to carry his values and the family values that he demonstrated throughout my life. Sometimes it's hard and life's challenges push us down different avenues, but if we can remain true to our values, we can do no harm." Payman looked the watch over as if it was for the first time. "Every time something good happens to me, part of me makes note of the watch and the man to whom it used to belong. I love it."

"As you should. It's a masterpiece." Maurice glanced down at his own Yacht Master that he awarded himself after his first admirable showing in America's Cup.

Maurice, more than almost anyone, knew the value of the programmable countdown and set the start of a regatta or set the start in life. As in any contest speed, proficiency, and precision are the essence of success.

"My grandfather had a real impact on me. I know that he sails with me every time I'm on the water and I look at his watch. It was his favorite possession and now it's mine. Family has such a profound, sometimes inexplicable impact on us."

"You're not wrong there," Maurice said simply. He had come from a land-loving, textile-producing family in the Pyrenees, and his father and grandfather saw in him the son that would carry on their family tradition of making some of the finest wools and winter weather clothing in Europe. However, he had different ideas and fell in love with the water from the babbling brooks that brought the snowfall down from the tops of the mountains, oftentimes canoeing and kayaking in the spring and summer snow melt. He fell in love with the water and followed it all the way to the sea where he found his passion and his true love in sailing on the open waters.

They both resumed their previous quietness, both men content.

"I suppose you've got something to brag to Bella about," he said. "I'm sure we'll all be telling our families of our victory."

"Lewis wouldn't," Payman said, then scrunched his face to hold back the laughter.

Their first day at sea was as it should be uneventful. Lewis had a chance to catch up on much-needed rest and was back to his normal self. Clint was in his comfort zone in the galley. John, content with the waves and the ship's steadiness, was primed to sit back and uncork a nice bottle of wine.

The sun didn't set early this time of year; there was nothing in the sky to give the crew an indication it was time for dinner. Their internal clocks, however, told them it was time to wind down the day.

Clint, with the abundance of Mediterranean seafood, prepared freshly caught seabass seasoned with fresh oregano, garlic, tomatoes, olives, and onions that he procured the day before. On the side, he prepared a penne pesto with fresh basil given to Lewis by Maria on his departure as one last well wish. One small gift from her garden that she hoped the guys would appreciate on their journey home. John delighted in serving the Benanti Etna Bianco, its crispness and fruity flavors perfect for the freshness of the fish and the basil. Clint and John toasted the crew for their excellent choices and their contributions.

Maurice broke the atmosphere with concern. "Even though the sea is perfect and the sky is clear, I hate to be the bearer of bad news, but we have a warning from the weather service of a severe storm forming into a medicane."

Lewis, ever the optimist, pointed to the sky and the horizon. "With these beautiful conditions, the chances had to be slight that such a storm would develop and even reach us."

The guys debated the threat that was posed in their imaginations and in the warnings of the weather service. Such a storm was unheard of in their memories. All agreed that such a storm was in their path and very intimidating on the open sea. But they held the optimism of all sailors and they knew they would be safe.

After the delicious Mediterranean fish, the crew readied the yacht for the worst. The air seemed heavier as the crew worked, the distant horizon cloaked in an unsettling stillness. Even as the sun dipped below the horizon, casting a fiery glow across the waves, a sense of foreboding lingered. Shadows stretched long and jagged over the deck, and the usual camaraderie among the crew softened to hushed tones and focused movements.

Maurice stood on the bow, his sharp gaze scanning the horizon. The water was eerily calm, no red sky at night, gave a sailor a warning, subtle tension hung in the air. Payman, ever the meticulous captain, gave orders as the crew lashed everything down, stowing their gear with precision honed by years of practice. Every rope was checked twice, every hatch secured, and every sail inspected. The metallic clang of tools and the shuffle of soles against the deck were the only sounds breaking the uneasy quiet.

Payman knew Maurice would ensure his orders were carried out, but he set about charting a route around the threat himself. The map before him felt like both a puzzle and a lifeline, the lines and coordinates offering the promise of safety if only he could navigate them just right.

As the last sliver of daylight disappeared, a cold wind whispered across the deck. Maurice called out, his voice steady but edged with concern. "We're ready, Captain. She's as tight as she'll ever be."

Payman nodded, his jaw set. "Good. Let's keep it that way."

That night, Maurice took the watch. The stars, normally a comforting sight, felt distant and cold. The sea, though smooth, seemed to hold its breath, each ripple carrying the weight of anticipation. Below deck, Payman settled into his bunk, the familiar creak of the boat around him oddly soothing. He ran through the checklist in his mind one last time, reassured by the crew's diligence.

They had a better team than he could've dreamed of—capable, disciplined, and united. They were prepared. They had a plan. They were going to be fine.

The water remained smooth, the sea breeze fresh, and the only sound was the craft slicing through the water. For a moment, Payman allowed himself to relax, letting the rhythm of the sea lull him. This was a day he would etch somewhere on his life's highlight reel. His family, his love for Bella, his crew, and his boat—all pieces of a life he wouldn't trade for anything. As sleep overtook him, a quiet prayer crossed his lips: gratitude for the blessings of the present and hope for the trials ahead.

CHAPTER 12
MEDICANE

Mid-Mediterranean 39.53931° N, 4.88135° E
11 September 04:45 Local Time

The Mediterranean, though a relatively small and enclosed sea, could unleash waves of exceptional ferocity, rising to heights of eight meters. These storms often created shorter distances between wave peaks, with severe angles and deep troughs—conditions extremely dangerous for any vessel. For a racing yacht designed for speed, it was a recipe for disaster.

Payman awoke abruptly to the sound of waves crashing against the bow, the noise reverberating just above his head. Disoriented, he lay in his bunk for a moment before the memories of the looming storm returned. He sprang to his feet, staggering slightly as the vessel shook violently. A quick glance out the small porthole confirmed his fears: towering waves rose and fell with relentless fury.

The deafening roar of thunder followed a web of lightning that scattered across the sky, illuminating the chaos outside. The ship jolted suddenly, throwing Payman against the edge of his bunk. Swearing under his breath, he scrambled to his feet, grabbed his heavy-weather gear from the wardrobe, and dashed to the deck to assess the situation.

Above, Maurice was shouting orders over the howling wind. The crew had steered the ship on an easterly path, hoping that as the storm moved northward along a westerly path they could miss the worst of the storm's fury. Even avoiding the worst of the storm, the outer bands were nonetheless relentless. The ship was either drenched in saltwater or about to take another wave-full. The unstoppable deluge of rain and waves made it hard even on the deck to hear his crew.

Payman pushed through the driving rain to join him at the helm.

"Skipper!" Maurice called, his voice barely audible over the cacophony. "I was hoping you could sleep through the worst of this!"

"Not a chance! The thunder and the waves crashing over my head made sure of that!"

Payman shouted back. "What's our status?"

Maurice pointed toward the horizon. "We're about 20 clicks from the heart of it! This isn't even the worst of the storm. The electrical activity is my biggest concern—we need to prep our electronics. Are you ready to take the helm?"

"I've got it!" Payman yelled, gripping the wheel. He turned to the crew and bellowed, "Everyone, get into your survival suits—now!"

The crew sprang into action, donning their immersion suits designed to prevent hypothermia and provide flotation in case of the worst. Payman tightened his grip on the helm as spider-like lightning illuminated the black sky, each strike briefly exposing the churning sea and the drenched crew.

Maurice's voice cut through the storm. "We need to decouple the electronics from the system! If we take a direct hit, we can't risk losing everything."

Three crew members moved swiftly to disconnect the equipment, their faces set with grim determination. Payman made sure the emergency line of-sight handheld radio was within reach, catching Maurice's eye and holding it up. Maurice nodded in understanding.

Without their weather radar, they relied on old-fashioned methods, tracking the lightning strikes and reading the storm's trajectory. Despite their efforts to steer clear, the relentless waves tossed the yacht as if it were a toy in a washing machine.

"We're in God's spin cycle!" Clint yelled, half-joking, though his voice betrayed his unease.

"What were the wind speeds before I took the helm?" Payman asked Maurice, his eyes scanning the horizon.

"Sixty knots, peak, maybe more!" Maurice shouted back.

Payman nodded and watched the waves, listened to the sound, and felt the wind. It was around 30 knots, 40 at most. The sea was still rising and falling. The ship was riding one swell after another. It felt like the bow and stern were rising at the same time, giving rise to wonder if there was any water below the keel. Then the bottom would drop like a roller coaster, in only enough time to catch a breath before the boat was tossed again in the next series of waves.

The yacht groaned under the strain as it rode each towering swell, the bow and stern rising and falling in a nauseating rhythm. Payman gritted his teeth, feeling the vessel shudder with every impact. The crew lashed themselves to the deck, their survival suits offering a modicum of security against the unforgiving sea.

Then, it happened. A blinding bolt of lightning struck the mast, and the force of the explosion threw the crew off their feet. Shouts filled the air as they scrambled to recover. Maurice's voice rang out. "Skipper! Are you okay? Where are you?"

A second lightning strike briefly illuminated the deck, revealing Payman crumpled against the port-side bulkhead, his limbs splayed awkwardly. "SKIPPER!" Lewis shouted, crawling toward him with cautious urgency. He knelt beside Payman, carefully assessing his condition before touching him. Blood seeped from Payman's lower body, staining his suit.

"Skipper, talk to me!" Lewis pleaded, cradling his head. Payman's eyes fluttered open, unfocused. Then, a cry of agony pierced through the storm's rage. John and Clint rushed over, their flashlights casting shaky beams of light on the scene.

Payman's leg was twisted unnaturally, the bone clearly broken. The crew worked quickly to stabilize him, constructing a makeshift tent to shield him from the relentless rain. Maurice, meanwhile, bolted to the emergency radio, his voice steady but urgent. "MAYDAY, MAYDAY! This is the 'Falkin. Can anyone hear me?"

The radio crackled before a deep voice replied. "'Falkin, this is the tanker Wassaja. We hear you. What's your status?"

"Wassaja, our captain is down with a severe leg injury," Maurice said. "We're battling a storm with eight souls on board. Our VHF radio is down. We need assistance!"

"Understood. We have you on radar, you're about a mile away. We're patching you through to the Italian Coast Guard. Stay on this line."

The seconds stretched into an eternity before the Coast Guard's voice cut through. "This is Captain Evangelista. We have your position and are dispatching medical and rescue teams. Hold tight, 'Falkin. Help is on the way."

Maurice exhaled sharply, his knuckles white on the radio. The crew exchanged grim but determined looks, rallying around their captain as they braced for the storm's final onslaught.

CHAPTER 13
TIB FIB

Mid-Mediterranean 39.53931° N, 4.88135° E
11 September 05:20 Local Time

The Italian Coast Guard, Corps of the Port Captains, is a distinguished division of the Italian Navy, whose mission is to protect navigation around the thousands of square miles of waters surrounding Italy's iconic "boot" homeland and its numerous islands. Renowned for their exceptional skills and professionalism, they are widely considered one of the best coast guards in the world. Captain Sal Evangelista, a seasoned veteran of the Coast Guard, was on duty when the distress call came in. The mayday signal from the 'Falkin echoed through the communications room, and Evangelista sprang into action. Quickly assessing the situation, he coordinated the response with practiced precision.

Understanding the gravity of the emergency, Evangelista connected Maurice from the 'Falkin with Dr. Matteo Zara, a navy lieutenant and emergency trauma doctor whose resume included service with the Italian Joint Special Forces and the Carabinieri Corps Special Unit. Matteo's calm voice cut through the chaos, guiding Maurice and the crew via radio through the critical steps of stabilizing Payman. While Captain Evangelista continued mobilizing every available resource for the rescue, Matteo's steady, self-assured bedside manner served as an emotional anchor for the battered vessel. Nature's fury was beyond their control—lightning lit the skies with blinding intensity, and the tenacious waves threw the yacht into a relentless rhythm—but Matteo's calm, confident tone reassured the crew that help was coming.

Within minutes, Maurice had fashioned a tourniquet above Payman's wound, started a saline drip, and stabilized him as best as possible under Matteo's direction. Even as the wind howled and the waves battered the boat, the crew clung to Matteo's assurances: Payman was stable, and rescue was on the way.

The storm's fury continued unabated. Payman remained in shock, his body spasming with pain despite the codeine administered through the saline drip. Lewis, crouched by his side, kept him warm with dry blankets while whispering reassurances over the storm's roar. Every wave that slammed the deck sent vibrations through the ship, causing the injured captain to writhe further in agony. The open tib-fib fracture posed a grave risk—sepsis, amputation, or worse—but the crew's relentless focus on his survival was a testament to their solidarity and resolve.

Amid the chaos, prayers to God were muttered in multiple languages, blending with the storm's cacophony. The crew, though battered and bruised, worked tirelessly to secure the ship and ensure Payman's comfort. One crew member, staggering past Lewis with a length of rope, shouted, "Where the hell is the rescue?!"

"It's coming!" Lewis shouted back. "Just keep us afloat!"

Maurice, hunched over the radio, cursed under his breath as the sway of the boat made it nearly impossible to steady his fingers.

The normally romantic sight of the endless sea now felt like a cruel void, mocking their desperation. He scanned the horizon, silently willing a silhouette—any sign of help—to break through the haze of rain and crashing waves. The crew adjusted their course to minimize the boat's movement, every action aimed at protecting their injured captain.

Cushions surrounded Payman to prevent further jolts, and Maurice repeatedly called for updates from the Coast Guard. The sea, though slightly calmer, remained an unforgiving expanse. Lewis kept a steady hand on Payman's pulse, whispering, "Hold on, Skipper. Help is on the way."

Then, finally, a crackle over the radio: "This is the Coast Guard, Captain Evangelista. We have you on radar and are dispatching medical and rescue teams to your location. Stay on this line."

The crew erupted in a collective sigh of relief, though their exhaustion was unmistakable. Evangelista's firm, composed voice brought a glimmer of hope to the battered sailors. As they continued their fight against the waves, Maurice shouted into the radio, "Thank God! Our captain is in critical condition—we need immediate evacuation!"

Evangelista replied, "Understood, 'Falkin. Stay steady. Medical assistance is en route." And then, through the haze of rain and despair, the flutter of helicopter blades whipped the air coupled with a faint silhouette appeared on the horizon. "I see them!" a crew member yelled, his voice cracking with emotion. A wave of renewed energy surged through the crew as they worked to prepare for the rescue. The vast, indifferent sea no longer felt so isolating; hope had arrived.

CHAPTER 14
GUARDIAN ANGELS

Mid-Mediterranean 40.08181° N, 5.28011° E

11 September 05:30 Local Time

While the crew on the 'Falkin fought to maintain hope, Captain Evangelista scoured for resources made scarce by the ongoing storm. All rescue teams were engaged as they battled the relentless elements. Operating under near-impossible conditions. Evangelista had connected Maurice with Dr. Zara, who calmly guided the crew through the delicate process of stabilizing Payman amidst the chaos. Despite their efforts, the Medicane raged on, consuming the region's resources and hope.

Unlike Dr. Zara, who had managed to guide Maurice in stabilizing the patient amidst the chaos, Evangelista found his efforts thwarted by the sheer magnitude of the storm's impact. The Medicane ravaged the Italian coast, leaving a trail of destruction in its wake. With all coastguard assets stretched thin. Every rescue boat, helicopter, and emergency team member were deployed in a desperate bid to save lives and mitigate damage, yet communications crackled with urgent calls for assistance and despite his best efforts, Evangelista's mission to assign resources to the 'Falkin rescue had been hampered.

Quantum

The west coast of Italy, along with the scattered islands of the Mediterranean, bore the brunt of the Medicane's wrath. Torrential rain and howling winds showed no sign of abating, transforming tranquil waters into a furious mass. The seas churned violently, waves reached towering heights, swallowing everything in their path and making navigation treacherous. Coastal towns and villages were inundated, and the relentless downpour threatened to erode the very foundations of the trust in the coastal infrastructure. The relentless storm tested his crew's endurance and resilience as the Italian communities rallied together to face the onslaught with whatever means they had.

Evangelista, amid the tempest, remained resolute, driven by an unshakable instinct to assist those in peril. With the Mediterranean storm consuming both resources and hope, his only reassurance came in the form of a multinational joint forces training exercise being conducted from the Aeronautica Militare Base in Sassari, Sardinia. Among the participating units was the elite U.S. Air Force 563rd Rescue Group out of Davis-Monthan Air Force Base in Tucson, Arizona. These were the tip-of-the-spear warriors of Combat Search and Rescue (CSAR), the elite of the elite.

When the emergency beacon went active, the 563rd sprang into action. The mission was immediately designated as a real-world launch. Aboard their HH-60W Jolly Green II helicopters, members of the 55th Rescue Squadron, along with 48th Rescue Squadron Guardian Angel teams composed of Pararescuemen (PJs) and Combat Rescue Officers, were scrambled. They moved with the fluid precision of warriors who had trained for moments just like this.

These are the airmen who go into denied territory to extract downed pilots and isolated personnel when no one else can reach them. Payman's condition met every criteria: time-critical, inaccessible by conventional rescue, and in hostile elements. The call was theirs.

The PJ Creed echoed in their minds: "It is my duty as a Pararescueman to save life and to aid the injured. I will be prepared at all times to perform my assigned duties quickly and efficiently, placing these duties before personal desires and comforts. These things we do, that others may live."

Becoming a PJ is no ordinary path. The pipeline, informally called "Superman School," is nearly two years long, featuring one of the most grueling special operations training regimens in the world. With an attrition rate around 80%, only the most resilient and capable earn the maroon beret.

When civilians think of the U.S. Air Force, they picture stealth bombers, fighter jets, satellites, and space operations. But it's through its airmen—the Guardians of the human domain—that the Air Force projects strength where it's needed most.

The 563rd received the OPORD (a formal directive to launch the real-world rescue op, detailing how to proceed tactically and logistically): a sailor aboard a racing yacht en route from Sardinia to Plymouth had sustained a traumatic leg injury. The vessel was caught in gale-force conditions, sending out a distress call that barely pierced the storm. With conditions deteriorating by the minute, the Rescue Team was wheels up within the hour.

The call came quick: launch one 79th Rescue Squadron HC-130J Combat King II for aerial refueling and comms relay, and two HH-60Ws for the insert. The PJs were briefed: they might have to perform a water jump and swim to the vessel. Either way, they were going in.

Colonel "Gandhi" Simpson, seasoned CSAR pilot and commander, took the stick on the lead HH-60W. His voice was a calm anchor in the chaos, revered by those who flew with him. He reviewed storm patterns and knew they had limited options. A direct fast rope insertion would be suicide with the yacht bucking like a bronco on steroids.

The crews lifted off in formation, rain hammering the fuselage and crosswinds buffeting the HH-60W Jolly Green IIs. Inside, blue-green cabin lights cast an eerie glow as turbulence rattled every bolt and every jaw. But under Gandhi's steady voice over comms, the crews held firm, riding the chaos like seasoned warfighters. Below them, the sea raged—whitecaps breaking over rolling swells like detonations in a minefield. Adrenaline surged. They weren't just on mission—they were on the clock. This was a real-world exfil. Lives were on the line, and the Guardian Angels were inbound.

In the cockpit, Gandhi studied the heaving ocean below. "Negative on fast rope," he radioed. "That deck's a death trap in this swell. We go wet."

Chief PJ "Gamer" Kent nodded, already prepping the team. "Copy that. We'll jump into the drink and swim to the target. Eight-footers or not, we'll make it."

As they neared the objective, the PJs locked into mission mode. Each operator ran through final checks—double- and triple-confirming their rigging, tightening chest harnesses, securing gear loops, verifying comms, and confirming lifelines were clipped and clear. Gloves were cinched, and NVGs packed and secured—

daylight was breaking. So they were unneeded for, but carried in case things went sideways. The cabin buzzed with quiet intensity—no words, just purpose.

Water insertions were notoriously dangerous. Known in the field as 'splash in' jumping into a dynamic, turbulent sea meant high risk of injury or separation. The sea state below was a frothing mess of whitecaps and rogue swells—zero margin for error.

Every second counted. One missed call, one bad timing on the wave trough, and it could go sideways fast. But these were Guardian Angels—trained for chaos, baptized by storm.

They were ready.

Gandhi held position over the swell, keeping the bird steady despite nature's fury. "Holding us as low as possible over the crest, say 5-10 feet and we'll have waves crest into the cabin and cover the tail wheel. Then you know we're good!," he called. "Timing's everything—we drop wrong, we're looking at broken bones or worse."

"Roger that," Gamer replied, steady as ever.

Final gear checks. Comms tight. Hearts calm. The storm surged. Lightning lit the sky as the first PJ leapt, followed by the others. Cold water slammed into them like concrete. Gamer barked, "No breaks. Head to the boat. Let's go!"

Aboard the 'Falkin, the crew held their breath as the whup-whup-whup of rotors grew louder. The VHF crackled, and Gandhi's voice came through: "the PJs are dropping into the water. Assist them aboard, ready the winch, follow orders precisely."

Relief swept over Maurice and the crew. They were ready to do whatever was necessary to get Payman to safety.

Acknowledging the instructions, Maurice turned his attention to the rescuers as they descended from the helicopter, dropping into the water just feet away.

Payman's condition continued to deteriorate. Maurice's chest tightened as he watched Lewis replace a blood-soaked bandage on Payman's wound. Through the mist, he could see Payman's body lying still—too still. The writhing agony that had marked the earlier hours was gone, replaced by a worrying lack of response.

Maurice squinted through the haze, his breath catching as he spotted the rescuers in the water, their silhouettes growing larger as they swam toward the boat.

"Come on, Skipper," Maurice muttered under his breath, his hands gripping the railing. "Just hold on a little longer."

Gamer and his team hit the water hard! The sea was a combatant in its own right, throwing up six-to-eight-foot swells that slammed into them like breaching charges. But this was not their first blue-water insertion. Instinct kicked in. Training took over.

They swam in tight formation, spacing kept deliberate to avoid collision but close enough for mutual support. Every stroke was synchronized with the rhythm of the sea, eyes locked on the pitching hull of the 'Falkin as it rose and fell like a bucking stallion. Gamer called cadence through short, sharp bursts over his waterproof comms, keeping the team aligned.

Each wave crashed down upon them like a rockslide. Their training took over, and their actions became automatic. As they neared the vessel, the yacht pitched and rolled wildly, challenging their every move. They timed their approach with precision, they synced their strokes to the chaotic rise and fall of the 'Falkin.

The PJs pushed through the rough seas. Though the rain and wind had calmed, the waves were still dangerously active. Finally, they reached the 'Falkin. Timing was everything. One mistimed grab at the ladder, and you risked getting crushed between hull and wave.

Onboard the 'Falkin the visibility was limited, the sea still angry but slowly calming. Maurice squinted through the mist, his breath catching as he spotted the swimmers in the water, their silhouettes growing larger as they swam toward the boat. The swimmers, clad in bright orange dry suits, stood out visibly against the dark waters. Though the storm cell had passed, the residual sea state was still hostile—whitecaps breaking over their heads, chop slapping against their faces. Their silhouettes growing larger as they swam toward the boat. As they approached the target vessel, they shifted to tactical float mode—adjusting posture to stay stable in the chop while preparing for boarding. One by one, they began climbing up the ladder, their movements quick but purposeful.

The crew rushed to the rail, their faces etched with a mix of hope and desperation. One by one, the PJs were hauled aboard—soaked, silent, and all business.

Maurice and the crew, their faces etched with anxiety and hope, the rescuers were aboard with a mix of relief and desperation. Gamer's presence was an immediate anchor in the storm, his calm authority cutting through the chaos like a lighthouse beam in the dark.

Gamer's presence hit like a command post beacon — cool, controlled, and mission-first. In the chaos, he brought the calm.

Maurice ran to meet them. "Thank God you're here!"

"Where's the injured?" Gamer shouted, wasting no time.

"Port side, mid-ship. The winch is in the rear."

The PJ crew, running on adrenaline and resolve, moved with swift urgency. Two rescuers rushed to Payman's side, while the others hauled gear aboard. Working in synchronized rhythm, the rescuers communicated in rapid, coded language, almost foreign to the crew. They quickly assessed Payman's condition, their eyes scanning every critical detail. They checked vitals, monitored the IV, and noted that Payman, though conscious, was in excruciating pain. They calmed him and administered another sedative. Inside the yacht, a new sense of determination took hold.

Under Gamer's guidance, they began preparing Payman for extraction. Every task was executed with precision, every movement calculated to minimize further injury. Above them, the Pave Hawk hovered—a lifeline in the chaos—its blades slicing through the air. Time was their enemy; every second stretched unbearably

As the rescue progressed, Gandhi monitored the situation from the helicopter, his hawk-like vigilance unwavering. Fighting the storm's relentless push and pull, he ensured the operation stayed on track. The combined determination of the 355th CSAR Group and the 'Falkin crew carved a path through the chaos.

Gamer called through his comms to the helicopter, "We're ready to go. Drop the cable slowly with the stokes-litter and a tag line to keep the litter from spinning. Let's get this guy up. Then we'll get the hook back and the rest of us will get out of here."

The litter descended from the hovering HH-60W with precision, its descent as deliberate as pouring a glass of Guinness, swinging slightly in the rotor wash like a pendulum on a timer. On deck, Gamer and his team moved with silent urgency. One PJ acted as the primary hook man, guiding the steel cable with gloved hands while maintaining tight comms with the flight engineer above. As the litter touched down on the deck with a metallic clank, they moved fast. It was released and positioned to secure Payman. Gandhi kept the bird rock steady, his hands locked on the cyclic and collective, compensating for crosswinds and yaw drift as the aircraft hovered in a tight footprint above the yacht's stern.

The PJs snapped into action. Gamer checked Payman's vitals one more time, then took point on securing him inside the litter, strapping him in with a tri-strap harness system to immobilize his injured leg, cinching restraints tight across his chest and thighs. A second PJ stabilized his head and shoulders, maintaining spinal alignment. Payman's face twisted in pain, but he was conscious—barely.

Once Payman was locked in and ready for vertical lift, the PJs flashed a "ready for hoist" hand signal to the crew above. Gamer clipped himself into the secondary D-ring beside the litter. The flight engineer confirmed over comms and initiated the lift.

The cable tightened, lifting slowly—deliberate and controlled to avoid swing. The stokes litter rose off the deck, swaying gently, with Gamer braced against the side. The tag line attached was managed by a PJ on the deck keeping tension on the line to keep the litter from swinging/spinning. They ascended in silence, rotor wash kicking spray in all directions. Up in the bird, the crew chief leaned out, guiding the litter in, hand-over-hand until the payload was secure inside the cabin.

Lewis watched Payman being received by the helicopter, then shared a silent look with Maurice. "He lost a lot of blood," Lewis said, his voice low. "I don't know if—" "You did everything right," Maurice interrupted, placing a hand on Lewis's shoulder. "It's out of our hands now. We've done all we could. Now let's get him to shore safely." At that moment, as if nature herself sought to grant them reprieve, the sun broke through the clouds, and the wind diminished.

Back on deck, the remaining PJs had already staged the exfil gear and cleared the boat deck for the second lift. The gear was next, vanishing like a magician's trick into the helicopter. The hook was dropped again and received. The remaining two PJs attached themselves to the cable. They clipped to the harnesses, and locked in with a tug and nod. The cable went taut again, and the final lift began. One PJ, casually waving as he prepared to rise, called out, "See you on stable ground." With that, the two PJ's ascended, rising slowly, silhouetted against the gray horizon, entered the open cabin door of the Jolly Green like shadows pulled from the sea.

The Jolly Green flew toward safety, leaving behind a battered but resilient crew. The 'Falkin crew watched as their friend was whisked away, their hearts heavy. They had faced the tempest and emerged with their spirits unbroken, their friend's life saved by the courage and skill of the rescuers.

CHAPTER 15
SASSARI

Sassari, Italy

11 September 08:30 Local Time

Following triage aboard the 'Falkin and a textbook hoist exfil, he was loaded the waiting HH-60W Jolly Green II from the USAF 563rd Rescue Group—a specialized CSAR (Combat Search and Rescue) unit with decades of dust-off legacy. The bird launched with urgency, en route to a Level I trauma facility in the coastal city of Sassari.

This wasn't your average medevac. The Jolly Green was rigged for both war and life—bristling with GAU-21s, countermeasure pods, and a full MEDEVAC suite onboard. Inside, the PJs worked efficiently, administering fluids, managing pain, and maintaining airway integrity. Payman, though sedated, was aware enough to know he was in the hands of the best.

As the chopper broke over the city's edge, its rotor wash stirred chaos on the rooftop landing zone. The LZ (Landing Zone) had been secured by local security forces, but the hospital staff had never seen anything like this—a combat-configured bird screaming in low and fast, flanked by the unmistakable profile of Guardian Angel teams in full tactical kit.

Gandhi called out the final approach over the intercom. "LZ in sight. Hot load, stay frosty."

The HH-60W touched down hard and fast. Crew chiefs sprang into action, sliding open the side doors. The senior PJ made first contact with the awaiting trauma team, giving a lightning-fast 9-line MEDEVAC handoff and confirming vitals, treatment administered, and potential complications.

To the civilian trauma surgeons, it was immediately evident: this wasn't a simple medevac. It was a Tier-One combat rescue, executed with surgical precision.

They moved fast. The litter was transferred, and the Guardian Angels stepped back, hands still on their gear, eyes scanning the perimeter by reflex.

"We've got him," the trauma lead shouted over the thumping rotors. "Thanks for bringing him home."

The PJs gave a silent nod and backed off, their mission complete. Payman was finally feet dry—safely back on land after a high-stakes extraction.

He was rushed into the trauma bay, where orthopedic and vascular surgeons began work. The damage to his leg was extensive—compound fractures to both the tibia and fibula—but thanks to rapid evac and flawless TCCC (Tactical Combat Casualty Care), the limb was viable.

Back on the rooftop, the HH-60W lifted off in a spray of dust and noise. Its silhouette faded into the dusk, leaving behind the distinct sound of blades cutting through air and the quiet reassurance that no one gets left behind.

After hours in the OR, titanium hardware was installed to stabilize the bones, align the tibia and fibula bones of his leg and secure them.

He'd forever carry metal inside him—setting off airport magnetometers for the rest of his life—but he'd walk again. And he was alive.

Later that day the crew had returned the 'Falkin to the harbor of Sassari. Though they were the same men who had celebrated their Rolex Cup victory just days before, they were now profoundly changed, deeply affected by the harrowing ordeal. Thankfully, no other injuries required medical attention, and all onboard systems were fully operational once again.

Bella had been notified, and she was on her way to Payman's side. Meanwhile, Lewis reached out to Maria and, to her surprise, invited her to Sassari—a short drive across the island from Porto Cervo. She was excited to see Lewis again but saddened by the news of Payman's injury. Lewis informed her that Payman's girlfriend, Bella, was en route from the UK to be by his side, suggesting it would be best for Maria to come alone.

Payman would not be ready to receive visitors until the next day, as the medical staff insisted he rest after surgery and recover from his concussion.

The time gave Lewis and Maurice an opportunity to explore the harbor city. Sassari's city center, centuries old yet beautifully preserved, was perched on a stone rise overlooking the surrounding area. Wandering its narrow streets and alleys, the pair found themselves lost in thought, their emotions still raw from the storm. Despite their somber mood, the Italians refused to let melancholy settle.

The owner of a small streetside trattoria, spotting the pair, called out in English, "Two such sad Americans will not do well for foreign relations!"

Maurice, switching to French, retorted, "I'm French, and he's English. Have some respect!"

"Ah, my apologies!" the man exclaimed, feigning shock. "Fellow Europeans, then. In that case, let me offer you a Prosecco as an apology. No need to be sad in the most beautiful place in the world. I'm Mario, and my trattoria is where you'll experience the best of our island."

Before they could respond, Mario beckoned to his barman for a bottle of Prosecco and three glasses. The trio sat at a patio table adorned with vibrant bougainvillea, the scent of the sea mingling with the mouthwatering aromas wafting from Mario's kitchen. The Prosecco, crisp and fruity, took the edge off their tension, easing them into the moment.

"So," Mario began, pouring the sparkling wine, "what brings you to our beautiful city? Perhaps the allure of Sassari's famous women? My wife is from here—after I met her, I couldn't leave!"

Lewis smiled faintly. "We're here under sadder circumstances. We're from the sailing ship 'Falkin. We won the Rolex Cup just four days ago."

Mario's eyes lit up. "A victory! What could possibly be sad, then?"

Lewis hesitated before continuing. "We got caught in the worst of the Medicane. Our boat was tossed like a toy. Lightning struck our mast, injuring several of us. Our skipper... he suffered a life-threatening leg fracture and a concussion. He had to be airlifted to the hospital. He's recovering now, but we can't visit him until tomorrow."

Mario's face softened. "I heard about that rescue during the storm. Our Italian Coast Guard—Corps of the Port Captains—is the best in the world. I'm glad they could save him. By the time this story reaches every bar in Sassari, your skipper will be a legend. When he recovers, he must come to my trattoria for dinner!"

Maurice smiled. "Thank you for your hospitality, and for joining us for a drink. Let us treat you to one in return."

Mario laughed. "Of course! You don't become the best trattoria in Sassari by staying in the kitchen. My job is to make sure my guests enjoy themselves."

Another bottle of Prosecco arrived as if by magic, loosening the tension in Lewis and Maurice for the first time in days. The three men talked like old friends, Mario's charm working its magic.

Eventually, Mario stood. "It's been a pleasure, my friends, but I must get to the kitchen and prepare your dinner."

Lewis grinned. "Thank you, Mario. This has been wonderful. I'm starving now—can we see a menu?"

Mario waved him off. "No, no! Tonight, I'll prepare something special. Antipasto, soup, salad, pasta, and porceddu—our famous roasted suckling pig. Save room for every course. It will be the best meal of your life!"

Lewis, always the adventurer and with the help of the Prosecco was all in for a meal and settled in for the duration. But first, “Mario, my beloved is joining me tonight.

“No problem there will be enough for three and I have an endless supply of Prosecco.”

“No I mean, we will be in need of romantic Auberge for several days, can you make a recommendation?”

“My friend, Lewis, I will arrange the best room at Mason De Charme” It’s a close walk. My sister owns it, it’s the best in the Sardinia. Please have you your lover join you here for dinner then you can go to the Auberge to sleep and enjoy you night with her on solid ground. We are a city on a rock. Sasso means stone. So Sassari literally means a city on stone. Enjoy my friends”

As the antipasto arrived, Lewis leaned back in his chair, finally allowing himself to relax. The storm felt like a distant memory, replaced by the warmth of Mario’s hospitality and the promise of better days.

CHAPTER 16
REUNION

Sassari, Italy

11 September 17:30 Local Time

Lewis reached out to Maria by text, sending her directions to the reserved parking spot Mario had graciously set aside for her behind the trattoria. The spot, tucked away for privacy, offered ease of access. Maria was only an hour's drive across the island in Porto Cervo, making it convenient for her to join them in Sassari. Despite the short drive, Lewis resisted the urge to bombard her with messages, though he couldn't help but send a few affectionate texts and emojis, telling her how much he was looking forward to seeing her. Maurice, ever the practical one, eventually made him stop, reminding him to let her drive safely.

As they waited for Maria's arrival, Lewis and Maurice settled into the trattoria's cozy warmth. The rustic charm of the stone walls and wooden beams exuded a sense of timeless tranquility, a stark contrast to the chaos they had endured. Outside, the Sardinian sunset cast a golden glow over the hills, and the soft murmur of conversations blended with the scent of fresh herbs and roasting meats from the kitchen.

The antipasto spread was a vibrant feast: marinated vegetables, cured meats, and artisanal cheeses artfully arranged. The deep greens of olives and roasted reds of peppers invited indulgence. As they savored each bite, the calming atmosphere began to settle their nerves, a stark reminder of how far they'd come from the storm's fury.

Lewis glanced at his phone, eager but resisting the urge to text again. Maurice, sensing his restlessness, smiled and raised his glass. "She'll be here soon. Let's toast—to survival and the people who make it worthwhile."

As their glasses clinked, the trattoria's warm lights flickered on, complementing the evening's serene ambiance. They were safe, Payman was recovering, and soon Maria would join them, completing the circle of resilience they had built.

When Maria arrived, Lewis nearly leaped from his seat. Their embrace was immediate, a melting of distance and worry.

"I've missed you these past two days," he murmured before their kiss deepened. Heads turned, some guests smiling warmly while others blushed.

One older Italian man chuckled, murmuring, "Oh quanto è meraviglioso l'amore."

Mario appeared, beaming. "Yes, how wonderful love is!"

"Oh, quanto sei bella," Oh how beautiful you are! Lewis proclaimed. Kissing her cheeks. She blushed but smiled brightly, ever the extrovert. "Sì, l'amore è italiano!" Yes, love is Italian, she declared, prompting glasses to rise in a chorus of "Saluti!" as the room celebrated their reunion.

Maurice greeted her with the European custom of cheek kisses. "You certainly know how to make an entrance—and a stunning one at that."

Maria, settling into her seat, asked, "How's the skipper? I've been worried."

Lewis's smile softened. "He's recovering. Surgery went well, though he'll be setting off airport metal detectors for life. We're lucky he's alive, but it was touch-and-go for a while. By God's will, we're all here, and he's on the mend."

Mario arrived and the two Italians dove into an emotional, animated discussion with their voices raised. They then abruptly smiled and hugged.

Lewis chuckled. "What was all that about?"

"Oh, I just told him how amazing his bistro is, how I love its atmosphere, how I'm so happy to see the both of you. I thanked him for the parking spot. We discussed the restaurant business as I spoke of being raised in the kitchen of my family's restaurant. He expressed his concern about the Skipper. And he said I have a surprise waiting for me after dinner. What is he talking about?"

"Oh, he has set us up with a suite at his sister's hotel. I'm sure that's what he's referring to."

"Isn't that presumptuous? How do you know you won't be spending the night there alone and I'll retreat to my own suite?"

Lewis felt a punch to his gut. The wind was knocked out of him like a sail softening with the doldrums. Lewis, forgetting Maurice was there, said "Oh I just thought after our amazing night together we could renew that experience." His eyes took on the appearance of puppy dog sadness. She reached over and kissed him. "I was only joking. I can't wait to spend all the time we have together."

Lewis responded first, "I was so sad and convinced that I'd have to keep Maurice company."

"Maybe you don't know what you're missing," Maurice said dryly before three of them laughed.

Their conversation turned lighter as Mario brought their next course: a rich seafood broth, fragrant with saffron and lobster. Maurice, savoring each spoonful, admitted it rivaled the best he'd had in France. Just as the soup was cleared, Mario appeared with wine and a playful announcement of the night's main course: Porceddu, Sardinia's celebrated roasted suckling pig. The aromas were intoxicating, and Maria's eyes lit up.

Instead of engaging the entre, Maria bent to Lewis' ear. She whispered, "I think I'm falling for you. I thought so when I met you, but tonight it's different. Be prepared for a night filled with me devouring your body and enjoying your pleasure. My mouth has more desire for you than any meal. Take your time and enjoy every second of this as I will enjoy every second of you." She finished with a little lick of her tongue around his ear.

As they ate, all Lewis could think of was Maria, he made efforts to keep himself calm as she comfortably conversed with him, Maurice, and the staff. She was whisked off to the kitchen to meet the back-of-the-house team.

Maurice commented, "You have hit the jackpot, my friend. If only you were French, you would know how to love a woman like Maria."

Returning to the table, she said, raising her glass, "This is a celebration to survival and the bonds forged in storms."

Later, Lewis and Maria strolled arm-in-arm toward the Maison de Charme, the cool evening air wrapping around them like a soft embrace. The hotel, a historic building adorned with balconies and lush vegetation, looked as though it had grown organically from the surrounding neighborhood. The night manager greeted them warmly, handing over keys and escorting them to a top-floor suite. Inside, fresh flowers and a chilled bottle of Prosecco awaited, adding a touch of elegance and romance.

Before the door had fully closed, they were in each other's arms. The urgency of their reunion was palpable, their kisses deep and searching, as though trying to make up for lost time. They undressed one another with a mix of fervor and tenderness, their hands exploring familiar curves and muscles.

Maria whispered, her voice husky with emotion, "I've missed you more than I can put into words."

Lewis smiled, pulling her close. "I feel like I've been waiting an eternity for this moment."

As their passion unfolded, every touch and movement spoke of their connection. They moved as one, savoring each other's presence with a depth that transcended words. When they finally collapsed onto the bed, spent but utterly content, they laughed softly, their foreheads resting together.

"I'm so glad we're here," Maria said, her voice barely above a whisper.

"Me too," Lewis replied, holding her close. "Tonight, there's nothing else in the world but us."

They drifted off to sleep in each other's arms, the trials of the past days fading into the background, replaced by the quiet joy of simply being together.

CHAPTER 17
RECOVERY

St. Centro Traumatologico e Ospedale Joseph
Sassari, Italy 13 - 15 September

The monitors cast an eerie green glow across the room, their steady beeping a soothing rhythm, like a slow metronome. The first person Payman saw was a doctor, patiently observing him as if expecting him to wake at that very moment. Her eyes brightened as he stirred.

"Welcome back," she said warmly. "You've had quite an experience."

Payman blinked a few times, his eyes adjusting to the light, before spotting the name pinned to her jacket. "Dr. Conti," he murmured, his voice still thick with sleep.

"You can call me Andrea," she replied with a smile.

He tried to process his surroundings, noticing the soft dusk light pouring in through the window at the end of the room. Slowly sitting upright, he stretched his stiff back. "That's a nice name," with a groggy voice he said.

"How are you feeling?" Andrea asked, observing him closely.

"Stiff," Payman replied, rolling his neck in the hopes of loosening his muscles.

Andrea nodded. "You're looking well, all things considered. You have a broken leg and a bump on the head, but you're on the mend. It seems you'll make a full recovery."

"Broken leg!? He bellowed as he tossed the sheet off to see his leg. "Oh shit there's a cast and a wire thing! Where am I?" he asked, glancing around.

"You're at our hospital in Sassari. Your friends have been very persistent, but I've kept them back to give you some time to rest. When you're ready, I'll have some food brought in."

Payman smiled faintly. "I hope this fine resort serves a proper English breakfast. The IV isn't going to cut it."

Andrea chuckled. "I assume your sense of humor is nothing new."

"I couldn't possibly say. Maybe ask my friends," he said.

Andrea grinned. "I'll see what's left in the kitchen, but just so you know—it's after seven in the evening."

As she left, the room fell quiet again. Payman tried to piece together his last memories: the storm, the 'Falkin, the crew, and the pounding rain. He had been fighting to keep them steady. His eyes drifted to his leg, and a sense of dread began to settle over him. When Andrea returned with food, he asked how his friends were doing. She assured him again that his crew was safe and waiting patiently to see him.

"They're fine," Andrea said. "I've insisted that you be given space and time to rest. From what I heard, your situation could've been a lot worse."

"It's okay," Payman said. "They're my crew. They can see me if they want." His eyes drooping.

"Right."

Payman picked at the meal—ham, eggs, and Italian cornettos filled with custard and chocolate—but his appetite waned as quickly as it had come. He did enjoy the cornettos. By the time the food was cleared, exhaustion overtook him once more, and he drifted into a restless sleep, accompanied by the steady cadence of the monitors.

Three days into his recovery, Payman woke with a mixed sense of relief and discomfort. The room was dimly lit, the soft glow of nightlights casting faint shadows across the medical equipment surrounding him. A digital clock on the bedside table displayed the time—it was nearly twelve hours later than when he had last drifted off. The faint hum of monitors filled the space, punctuated by the steady cadence of his heart monitor.

As he shifted slightly in the bed, a dull ache shot through his leg, a reminder of the surgery he'd undergone. Payman winced but allowed himself a moment of gratitude. He was alive. The storm, the chaos, and the rescue some memory was returning—it all felt like a surreal blur. Hunger gnawed at him as he glanced at the tray of cold hospital food beside his bed. The uninspiring offerings did little to entice him. With a sigh, he closed his eyes, imagining the taste of a full English breakfast—a comforting memory that brought a faint smile to his lips.

Determined to indulge in something more satisfying, Payman pressed the call button. A young hospital aide appeared moments later, cheerful and eager to assist. "Cornetti e cappuccini per favore" In perfect Italian, Payman said with a warm smile. "Please, if it's not too much trouble. I'd appreciate a small taste of the outside world."

The aide chuckled and disappeared. Not long after, he returned with two steaming cappuccinos and a plate of pastries. The rich aroma of coffee filled the air, mingling with the sweet scents of chocolate, custard, and strawberry. Payman beamed as he expressed his gratitude, savoring each bite and sip as warmth and comfort slowly replaced his lingering unease.

Just then, Dr. Andrea Conti entered, clipboard in hand. Her kind smile and calm demeanor immediately put Payman at ease. "Good evening, Captain," she said, taking a seat beside his bed. "How are you feeling?"

"Better, thanks to these," Payman gestured toward the cornettos and cappuccinos, "and the fine care you've given me."

Andrea chuckled softly. "I see you've rediscovered your humor. That's always a good sign. Let's take a look and see how you're doing. If all is well, I'll clear you for visitors."

Payman nodded eagerly, longing to reconnect with his crew and loved ones. As Andrea examined him, her tone remained light, but her words carried weight. "You've made remarkable progress, considering the circumstances. I can't overstate how critical your rescue was. Your leg," she gestured to the cast, "was just one of many concerns. Honestly, it's a miracle you're here."

Payman frowned slightly, trying to piece together the fragments of his memory. "The rescue team—they were incredible, weren't they?"

Andrea's eyes softened. "They were more than incredible. They were your Guardian Angels, in every sense of the word. Let me tell your friends you can have a visitor, one at a time."

Bella entered the room, her tearful eyes lighting up as they locked onto Payman's. She rushed to his bedside, her trembling hand reaching for his. "Oh, Payman," she whispered, her voice thick with emotion. "I was so scared. I thought I might lose you."

Payman squeezed her hand weakly. "I'm here, love. Thanks to them—and thanks to you."

Andrea discreetly stepped aside as Bella leaned closer, her tears mingling with laughter. "You're a stubborn one, aren't you?"

"I had to be," Payman replied. "I couldn't leave my crew—or you." He gestured toward the heart monitor, watching the steady green line. "Besides, this thing says I've got a few more beats left."

Bella chuckled through her tears, pressing her forehead against his. "I love you, Payman."

"I love you too," he whispered. "More than you know."

Moments later, Andrea returned. She had been overwhelmed by the other visitors and agreed they could enter. Lewis, Maria, and a stranger in a military uniform joined the room, their expressions a mix of relief and curiosity. Lewis wasted no time breaking the tension. "Captain, I thought you were supposed to go down with the ship, not the other way around."

"Next time, you're riding the bow in a storm," Payman quipped, a grin spreading across his face.

Each of them shared Bella's reaction—overwhelmed with relief and awe—but no one could quite manage a proper hug with Payman. He was still fragile, but the energy in the room pulsed with quiet joy.

Lewis was eager to recount the harrowing rescue but respectfully deferred, "I'll let the Colonel take it from here," he said, nodding toward Gandhi.

"Hey, Payman," Gandhi began, stepping closer with a calm, commanding presence. "We met briefly aboard the helicopter. I'm Colonel Tim Simpson, U.S. Air Force—call sign 'Gandhi.' First things first: I'm glad to see you're progressing with your recovery. Next, I know you're all wondering about the name. It's a call sign—most of them aren't flattering, but I lucked out. I got mine after an eye infection years ago. I wore these round, wire-rimmed glasses—I kinda liked them, actually—but my crew thought I looked like Mahatma Gandhi. Shaved head and all. Could've been worse. Some flyers get stuck with names like 'Pop-Tart', 'Crash' or 'Slider.'"

Everyone chuckled.

Payman shook his hand, his curiosity piqued. "Gandhi? What are you protesting these days?"

The room erupted in laughter as Gandhi explained.

"I was the mission commander for your rescue," he continued.

Payman, his voice gravelly but full of emotion, replied, "So you're the one who sent angels after me—after I'd already made peace with not surviving. Thank you. From the bottom of my heart. You're not like angels… you are truly Guardian Angels."

Gandhi gave a humble smile. "Not sure all my commanders would call me an angel. I've done my share of devilish things. But yeah—we're called Guardian Angels. We're the guys who go in when someone's in a situation they're not supposed to survive. That's our job."

He paused, then commented, "we make sure someone's worst day…is not their last day."

The room went still.

Then Gandhi chuckled to break the spell, "I love my job!" He began to tell the full story of the rescue, this time from the perspective of the commander. Even Lewis, who had lived it, listened in awe. The tension, the coordination, the decisions made under pressure—it all painted a vivid picture of what had truly transpired above and on the sea that day.

"Guardian Angels are pararescuemen—PJs," Gandhi explained. "They're the tip of the spear in Combat Search and Rescue. Most people think of the Air Force as jets, stealth bombers, satellites. But none of those mean a thing if you lose the human in the cockpit. That's where we come in. All the pilots know: if something goes wrong, we're the ones who bring them home."

Payman listened, deeply moved. He'd heard of the Guardian Angels before—through military briefings during his training in Iran, and in films he'd seen as a young man. But now, having lived through their mission firsthand, his gratitude overflowed.

"How could I ever thank you for saving my life? For being there when… when I thought it was the end?"

"There's nothing you need to do," Gandhi replied. "This is what we do. It's our mission, our purpose."

Payman nodded solemnly, then said, "Do you know the movie Behind Enemy Lines? It was one of the most intense films I've ever watched. I know that movie is about you folks. Let me see if I can recall, the story behind it—US Air Force Lt. Col. Dale Zelko—he flew an F-117 stealth fighter over Kosovo."

Gandhi smiled. "Yeah. I know that one well."

He went on to fill in the gaps in Payman's knowledge. Zelko had spotted two SAMs, surface to air missiles punching through cloud cover. The first missile missed. The second exploded nearby—shrapnel and shockwave tearing through the aircraft, forcing him to eject. Still descending by parachute, he called out on his survival radio to maximize signal reach and confirm he was still alive.

"He was a huge target," Gandhi explained. "A downed stealth pilot—every faction wanted to capture him. He landed about a mile from his crash site and took cover in a drainage ditch. Yugoslav troops, police, even civilians searched for him. At one point, they were just a few hundred meters away."

"Then came the Guardian Angels," Payman said, eyes wide.

"Exactly. They dropped in, surrounded him, and shielded him with their bodies. Hauled him into the helicopter. He said, he was terrified lying at the bottom of a pile of guys while bullets were flying from below." Payman laughed gently.

Gandhi continued, "He joked about it afterward—said, 'Thanks for putting me on the bottom!' One of the PJs told him, 'Relax. The floor's armored. The doors aren't. You were in the safest spot in the bird.'"

Everyone in the room smiled.

Payman sat back, emotion welling in his eyes. "That was my favorite warfighter movie. And now I've lived something like it. I heard Zelko sends a bottle of 30-year-old scotch to his Guardian Angel team every year."

"He does," Gandhi confirmed. "Without fail."

When Gandhi finished, Payman's voice was thick with emotion. "How can I ever repay you for what you've done?

I'm not much of a scotch drinker. But I do love bourbon. As long as I live, you and your team will get a barrel-aged bottle of Woodford Reserve every year. I hope your team puts it to good use."

Gandhi nodded, touched by the sentiment. "We will. And we'll raise a glass to you, every time."

"May God bless you all," Payman said, his voice steady.

CHAPTER 18
RELEASE

St. Centro Traumatologico e Ospedale Joseph
Sassari, Italy 16 September 08:20 Local Time

For the next two days, Bella remained by Payman's bedside, barely stepping away during her time in Sassari. The hospital, far removed from the luxury of the Four Seasons, offered little comfort. Still, Bella made the best of it, treasuring the moments as Payman gradually regained strength. He could now sit up, watch Italian TV, challenge her to spirited games of backgammon, and savor the delicious bites that Lewis and Maria brought from outside. While the others indulged in wine, Payman abstained, focusing on his recovery with a determination Bella found admirable. Yet the incessant beeping of monitors, the green fluorescent lighting, the chatter of Italian television, and Payman's thunderous snores during sleep began to weigh on her. She needed an escape.

"Maria," Bella said with a warm smile, "let's take a walk. I need fresh air and a change of scenery."

Maria's brow softened as she nodded. "Of course, Bella. Let's go."

The two women strolled into the heart of Sassari, their footsteps light against the cobblestone streets that wound through the old town. The streets bustled with life—vendors called out in melodic Italian, children's laughter rang through the air, and the warm afternoon sun painted everything in a golden glow. They stopped at a boutique filled with handmade goods and colorful fabrics. Maria admired a hand-stitched shawl while Bella fingered a delicate ceramic pendant. They laughed easily, their voices blending into the lively hum around them. It was easy being together.

They eventually found a small trattoria tucked away under a canopy of bright red bougainvillea in full bloom. The air smelled of fresh basil, sun-dried tomatoes, and the salty breeze from the nearby coast. The women settled at a table outside, and as they sipped crisp Vermentino and sampled pasta that seemed to melt in their mouths, their conversation deepened.

Bella leaned forward, her curiosity piqued. "Maria, I'd love to hear more about your family. It's obvious Lewis is smitten with you."

Bella had known Lewis for almost as long as she had known Payman. The three of them often spent weekends on the Plymouth and Firestone bays or sailed the breathtaking waters off the South Coast. Over time, Lewis became much like a brother to Bella. She trusted him implicitly, especially during challenging days at sea.

Despite Lewis's striking looks that attracted attention from women, Bella sensed a hint of loneliness in him. She knew he longed to find a life partner and settle down. Now, in a whirlwind turn of events, he radiated happiness whenever Maria was mentioned or in his presence. Bella had never seen him so content.

Maria chuckled softly, her cheeks flushing. "Well, my family has lived in our sea town for generations. We were one of the founding families of the yacht club that hosts the Rolex Cup every year. It's a chaotic but magical time. Normally, I'd be working at the family restaurant, but for the race, I always take time off. Being at the docks when the boats arrive—it's a tradition I treasure."

"Lewis raves about your restaurant," Bella said with a knowing smile. "And not just because he's with you."

Maria laughed, shaking her head. "I'll remind him of that later. But really, the first time I saw him…" Her voice softened. "It was on one of the early practice days. There was this sailor—handsome, yes, but more than that. He carried himself with such calm confidence, such warmth. I had to meet him."

Bella raised an eyebrow. "And?"

"I approached him at a reception for the racers and locals. He was everything I hoped for—gentlemanly, intelligent, genuine. By the end of the evening, I invited him to my second-favorite trattoria, and to my delight, he said, 'Let's go now before the crew ropes me into a meeting.'"

Bella laughed. "That's not like Lewis. Usually, he's the one doing the chasing."

Maria's expression softened as she continued. "On the final race day, I knew he was special. When his boat won, the whole dock erupted, but all I saw was him. He came straight to me, and when we hugged and kissed, I knew I didn't want to let him go."

Bella reached across the table, squeezing Maria's hand. "It sounds like a storybook romance, and yet here you are, living it."

Maria's smile faltered slightly. "When he sailed away, my heart broke. I finally understood what it's like for women of the sea to watch their men vanish over the horizon. Even when he called from the yacht, I couldn't shake my worry. When he told me they were heading into a storm, I prayed harder than I ever had."

Bella's voice quivered. "I didn't realize you were in touch with him during the storm. It must have been terrifying."

Maria nodded, her eyes glistening. "My mother always says it's our job to pray and send them our love. That's what Sardinian women have done for generations. But I was a wreck until I knew he was safe."

Bella's heart ached at Maria's words. "You've been through so much. No wonder Lewis is so devoted to you."

Maria's smile returned, this time stronger. "I don't ever want to see him sail away again."

Bella raised her glass. "Then let's toast—to the men we love and the strength it takes to hold onto them."

The clink of their glasses seemed to echo with promise, the warmth of their bond wrapping around them like the sunlit breeze. Bella felt a renewed certainty that their lives were inextricably tied—four hearts united by love, loyalty, and the unpredictable tides of life.

Good news arrived as Payman received clearance for release from the hospital. Despite still requiring crutches for walking, he was well on the path to recovery. Following a final checkup by his doctor and the necessary paperwork processing, Payman bid farewell to the hospital staff.

As the quartet exited the hospital, a sense of joy and camaraderie filled the air. Payman was comfortably seated in a wheelchair, helped by Lewis (although Lewis jokingly called it "Piloting Payman"), while Maria and Bella strolled gracefully down the hall, holding one another's hands. Indicating they had reached a new level of friendship following their girl talk yesterday.

The day unfolded beneath a canopy of brilliant sunshine, accompanied by a gentle breeze carrying the fragrant aroma of blossoming flowers and the invigorating scent of the nearby ocean. The surroundings seemed to come alive with vibrant colors and sounds, creating a picturesque backdrop for the group's shared journey toward brighter days ahead.

"Watch out, ladies and gentlemen, coming through! Special delivery for Mr. Payman," Lewis quipped, steering the wheelchair down the sloped hospital ramp with an exaggerated flourish.

Payman rolled his eyes, a grin tugging at his lips. "Just don't crash me into anything, Captain Lewis. I'd like to avoid coming back here so soon."

Bella chimed in, her laughter bubbling. "I think you're in good hands, Payman, but maybe we should have requested a seatbelt just in case."

Maria added, smirking, "Or a warning sign for reckless drivers. "Hey!" Lewis shot back with mock indignation. "I'll have you know I'm an excellent driver… on land and sea."

As they reached the bottom of the ramp, Payman stretched his arms theatrically. "Freedom! Sweet, sweet freedom. Now, what's the first order of business? I vote for real food. No more of that so-called 'hospital cuisine.'"

Bella leaned over, playfully tapping his shoulder. "You mean you didn't enjoy the gourmet gelatin cups and overcooked pasta?"

"Oh, absolutely. I even asked for the recipe so I can recreate that five-star rubbery delight at home." Payman retorted with a chuckle.

Maria's eyes twinkled as she suggested, "How about we head to the little café by the marina? The one with the fresh pastries and that heavenly espresso?"

"Now you're speaking my language," Payman agreed enthusiastically, as Lewis wheeled him toward the waiting car.

The group paused briefly at the hospital's entrance, taking in the brilliant sunshine and the lively hum of the city. Seagulls cried overhead, and the distant sound of waves crashing against the shore added a sense of calm. Bella inhaled deeply. "It feels like life is beginning again," she said softly.

Maria reached for her hand and squeezed it gently. "It is. For all of us."

Lewis, ever the joker, couldn't resist breaking the sentimental moment. "Alright, ladies, let's get moving. Payman's escape vehicle is double-parked, and I'm not paying a ticket."

With laughter ringing through the air, they piled into the car, the warmth of their friendship wrapping around them like the sunshine above. As they drove off, the promise of brighter days ahead felt closer than ever.

CHAPTER 19
BACK AT BENTLEY HALL

Bentley Hall Oxfordshire, England

24 November 14:30 Local Time

One rainy afternoon, as Payman recuperated in Iran from his harrowing brush with death, Bella and Bea found themselves immersed in the scholarly ambiance of the estate's library. The soft pattering of rain against the windows mingled with the scent of old books, creating an almost hypnotic atmosphere. Bella's gaze lingered on an imposing full sized ancestral portrait leaning casually against the wall. She stood to examine the portrait. It always intrigued her that her many generations back grandfather would be so vain to have a full sized painting done. But such were the times, for the rail barons. She examined the portrait. Her curiosity deepened when she noticed a wall of drawers concealed behind it.

"Bea," Bella called, her voice tinged with intrigue. "Come over here, girl. There's something behind this portrait."

Bea joined her, glancing skeptically at the long forgotten wall hidden by the portrait.

"You think it's a secret stash of treasure, or maybe just great-grandpa's old tax records?"

Bea smirked. "Only one way to find out."

Together, they carefully shifted the heavy portrait, mindful of the delicate flooring beneath. Dust swirled in the warm light streaming through the library's time-stained windows. What lay behind was a set of drawers, shrouded in years of neglect and cloaked in an almost sacred silence.

Bea brushed her fingers against the wood. "These drawers look like they've been untouched for centuries."

"After you," Bella said, gesturing toward them with exaggerated deference.

Bea raised an eyebrow. "What's this, suddenly losing your adventurous streak?"

"Not at all," Bella retorted with a grin. "I just thought you'd want first dibs. Unless, of course, you're scared of whatever might be lurking in there."

Bea chuckled, rolling up her sleeves. "Fine, but if I find cursed jewelry or an angry ghost, you're handling the cleanup. Anything valuable we split"

The drawers creaked open reluctantly, as if guarding their secrets even now. Inside were brittle pages, carefully folded maps, and documents chronicling generations of the estate's history. They sifted through birth records, debt tallies, and correspondence that painted a picture of the family's industrious rise—from managing river barges and canals to building railroads and, eventually, conquering the seas. Much of it nothing but brittle pages.

"Wait, I think I found something interesting. Listen to this," Bea said, reading aloud. "'Lord Andrew Bentley foresaw the decline of canal transport and championed railways. By 1833, he spearheaded the London-to-Penzance line, overcoming engineering challenges and financial skepticism. Ambitious guy."

"Ambitious and effective," Bella added. "This family wasn't afraid to adapt."

They delved deeper, uncovering tales of the Industrial Revolution and Railway Mania, where family ventures soared to prominence. Maps detailing railway expansions and financial ledgers told a story of both triumph and risk.

As Bea reached for the last drawer, her fingers brushed against the wood, and something shifted beneath her touch. A faint click reverberated through the room.

Bella froze. "What was that, the ghost?"

Bea looked up, wide-eyed. "I think… I think I just unlocked something."

The panel behind the drawers shifted slightly, revealing a sliver of darkness beyond. Bella and Bea exchanged a glance, their expressions a mixture of excitement and apprehension.

"You think it's a hidden passage?" Bella asked, her voice barely above a whisper.

"Maybe," Bea said, pulling out her phone and switching on the flashlight feature. "Could be a priest hole. Or a smuggler's tunnel."

"Or something worse," Bella teased, though her eyes sparkled with anticipation.

The passage was narrow and damp, the air cool and heavy with the scent of centuries-old stone and wood. Bella hesitated. “Should we tell Uncle Andrew? Let him know what we’re up to?”

Bea shook her head with a grin. “And ruin the mystery? Besides, he’s probably too busy making dinner or just directing the staff.”

Their footsteps echoed against the stone as they ventured deeper. Shadows danced on the walls, twisting and flickering with the light from Bea’s phone. The musty air seemed to press in around them, heightening the sense of suspense.

“Do you think this leads anywhere significant?” Bella asked, her voice tinged with awe.

“Only one way to find out,” Bea replied. “But I swear, if we find skeletons down here…”

As Bella and Bea ventured deeper into the narrow, hidden passageway, the walls grew colder and damper. The air was thick with the earthy scent of moss and stone, mixed with an underlying metallic tang that hinted at long-forgotten secrets. Their footsteps echoed faintly, the only sound in the suffocating silence of the corridor. The passage sloped slightly downward, its stone steps uneven from centuries of wear.

“Do you feel that?” Bella asked, her voice hushed, almost reverent.

Bea nodded. “It’s cooler down here, and the air feels… older. Like we’re walking into another time.”

As they pressed on, Bea’s flashlight beam illuminated faint carvings on the walls—symbols and etchings so weathered they were almost imperceptible. Some were indecipherable, but others appeared to be crude depictions of tools, water channels, and ships. Bella ran her fingers over one carving, her breath catching.

"These look like they were made by workers, maybe even builders from the time the estate was first constructed," she murmured. "It's like a story written in stone."

Bea crouched down to examine another carving, her curiosity piqued. "Or warnings," she said with a wry grin. "Don't tell me you're not getting a little creeped out."

Bella gave a nervous laugh. "Maybe just a little."

The passage eventually opened into a larger chamber, its vaulted ceiling supported by thick stone columns. The room was bathed in shadows, but the faintest glimmer of light filtered through narrow slits high in the walls. The source of the light was unclear, but it added an eerie glow to the room. At the center stood an ancient iron grate, partially rusted, covering what appeared to be a deep, dark well.

"What do you think this is?" Bea asked, shining her light on the grate.

Bella peered closer, noticing the faint smell of water rising from below. "A cistern, maybe? It could have been part of the estate's original water system. Look there's water below."

Bea's flashlight swept across the room, revealing more details—discarded tools, decayed wooden crates, and a set of stairs descending further into the darkness. A low archway at the far end of the chamber led to yet another passage.

"This place is like a labyrinth," Bea said. "Do you think anyone's been down here recently?"

Bella shook her head. "Not likely. If they had, there wouldn't be so much dust and debris. This has been forgotten for a long time."

The two exchanged a glance, excitement mingling with apprehension. Bea took the lead, heading toward the archway. As they passed through, the passage narrowed again, the walls pressing closer. The floor became rougher, and the air carried the faintest hint of metal.

"I think we're getting closer to the basement levels," Bella said. "Or maybe even a tunnel leading out to the river."

As they continued, they came upon an iron door, its surface speckled with rust but still intact. An ornate lock, long corroded, hung from the latch. Bea reached out to test the handle, and to her surprise, it creaked open with minimal resistance. Beyond the door lay a steep, spiraling staircase that descended into the depths.

"Do we go down?" Bea asked, glancing back at Bella.

Bella hesitated, then nodded. "We've come this far. Let's see where it leads."

The staircase wound downward for what felt like an eternity, the air growing colder and damper with each step. When they finally reached the bottom, they found themselves in a vast underground chamber. The walls were lined with alcoves, each containing objects shrouded in thick layers of dust—ancient tools, faded maps, and crates stamped with shipping logos from centuries past.

"This must have been part of the estate's trade network," Bella said, her voice tinged with awe. "A hidden storage area for goods being moved along the river."

Bea's flashlight caught something glinting in one of the alcoves. She stepped closer and uncovered a chest, its brass fittings tarnished but still gleaming faintly. A 'captain's chest', very old, maybe from one of the many family ships of the 18th or 19th centaury. She hesitated, then lifted the lid. Inside were stacks of papers, what appeared to be a ledger and a glass jar of coins.

"Jackpot," Bea said with a grin.

Bella leaned over to examine the contents. "These coins... Look at the markings—they're from all over the world. This must have keepsakes from ports visited as a part of the family's shipping empire."

As they sifted through the chest, Bella found a small, leather-bound journal tucked beneath the coins. She opened it carefully, the brittle pages crackling under her touch. The handwriting was neat and deliberate, its ink faded but still legible.

"It's a captain's log," Bella said, her eyes scanning the entries. "It details voyages, cargo, even dealings with foreign merchants. This could be invaluable."

Bea looked around the chamber, her imagination racing. "Do you think there's more like this? Other chambers, other records or even more treasures?"

Bella nodded, her heart pounding. "It's possible, who knows what else we might find?"

The prospect of uncovering more secrets spurred them onward, their excitement eclipsing their earlier apprehension. As they explored further, the passages revealed glimpses of a bygone era, their mysteries waiting patiently to be unraveled.

As Bella and Bea pressed on through the labyrinthine tunnels, their steps grew more purposeful. The air shifted slightly—warmer, with a faint trace of familiarity that Bella couldn't quite place. The walls of the passage seemed less damp, the stonework more refined, and the occasional wooden beam reinforced the ceiling.

"This feels… different," Bella murmured, glancing at Bea. "Not as ancient, maybe?"

Bea nodded, shining her flashlight ahead. "It's like we're coming back to a part of the house that's been used more recently."

The passage turned sharply, leading them to yet another iron door. Unlike the one before, this one was less corroded, its surface marked by faint scratches and worn edges. Bella's heart skipped a beat as recognition flickered in her mind. She reached out and ran her fingers along the door's cold surface.

"This door…" Bella said, her voice trailing off. "I think I know this door."

Bea raised an eyebrow. "What do you mean? Have you been here before?"

Bella hesitated, her hand lingering on the handle. "From the other side. When I was a kid, I used to sneak into the basement to explore. My uncle told me not to go too far, but there was this old door I was always too scared to open." She glanced at Bea with wide eyes. "I think this is that door."

"Well," Bea said, her tone light but laced with anticipation, "you're not a kid anymore. Want to see what's on the other side?"

With a deep breath, Bella turned the handle. The door creaked open, revealing a staircase that sloped gently upward. The familiar scent of the basement—earthy, with hints of cedar and the faint tang of stored wine—washed over Bella, bringing a flood of memories.

"It is the basement," Bella said, her voice soft with wonder. "I know this place."

They stepped through the door, and the narrow, damp passage gave way to the spacious cellar Bella had explored countless times as a child. The shelves along the walls were lined with dusty bottles of wine, jars of preserves, crates and old furniture covered in sheets. A small, dim light bulb flickered overhead, illuminating the space with a warm, golden glow.

Bea turned to Bella, a smile tugging at her lips. "Looks like your childhood curiosity was onto something. Who knew this basement was connected to all of that?"

Bella chuckled, her tension easing as the familiarity of the space settled over her. "I can't believe it. All those times I played down here, and I had no idea there was a whole hidden passage behind that door."

Bea leaned against one of the shelves, letting out a breath. "So, what now? Do we tell Uncle Andrew, or do we keep this little adventure to ourselves?"

Bella's gaze swept over the room, landing on a stack of old books she used to thumb through as a child. "Let's keep it to ourselves for now," she said with a mischievous smile. "I think this is just the beginning of what we'll find."

The two women shared a conspiratorial laugh, the tension of the hidden passage replaced by a shared sense of accomplishment and discovery. As they made their way up the familiar staircase that led to the main part of the house, Bella couldn't shake the feeling that the estate still had more secrets waiting to be uncovered—and she couldn't wait to find them.

CHAPTER 20
THE VILLAGE

Bentley Abbey Village Oxfordshire, England

25 November 17:10 Local Time

Below Bentley Hall, at the end of the gravel drive, nestled in the bend of the river near the old stone bridge, a quaint hamlet unfolded. Where there were once three taverns, only one remained. It was originally known as The Tipsy Friar, but long ago, out of respect for the local clergy, the name had been changed to The Globe after a large historic globe that sat prominently in the tavern's center. The sign announcing its location depicted a salty seafarer in a weathered coat and watch cap, charting his course on a globe. No one could say for sure how old the sign was, but some old-timers swore they remembered it from when they were in knickers— as they recalled, maybe 200, even 250 years ago! At this little corner of the Cotswolds, far from the sea, the dream of adventure lived on.

The Globe, with its honey-colored stone and thatched roof, fit perfectly with the surrounding cluster of cottages. Inside, low ceilings adorned with exposed timber beams gave the tavern a warm, inviting charm. Antique wooden furniture dotted the space, while a crackling fireplace cast a golden glow. The entrance door required most guests to stoop as they entered, the head jamb barely clearing 160 centimeters. Above the door, a cheeky sign read: "Too Late, Mind Your Head Next Time." At the tavern's heart stood the namesake globe, over a meter in diameter, mounted in an elegant and polished brass stand. It was a magnet for conversation, sparking tales of places people had been, dreamed of visiting, or simply boasted about in wild, fanciful stories.

The Globe's patrons ranged from weathered farmers swapping tales of their fields to jovial villagers deep in animated conversation. The clink of pint glasses mingled with hearty laughter. Both regulars and travelers alike found solace in the tavern's timeless charm, a gathering place where the world felt both vast and intimate.

This was a favorite haunt of the Bentley Hall trio. If it were up to Bea, they'd come here daily. As they approached, a chorus of greetings erupted from the locals.

"Oy, young Bella, daughter of beautiful Mary, watch your head, love!" shouted Old Harry, sitting by the fire with his ever-present walking stick.

"And keep an eye on Andrew!" added Martha, a sharp-tongued widow known for her endless stream of advice. "He's a thorn among you roses."

Bella ducked under the low doorway with practiced ease, flashing a grin at the room. "Evening, Martha! I'll keep Andrew in check, don't you worry."

"You'd better," Martha shot back. "If he gets up to trouble, I'll sort him out myself."

Another voice called from the bar. "The Bentleys are drinking for free tonight! Drinks on me!" It was young Tom, the village butcher, always eager to make an impression.

Arthur, the tavern's owner, let out a booming laugh. "That'll be the day, Tom! You've got a tab longer than the river. Keep dreaming, lad." He turned his attention to Bella, his expression softening. "What'll it be, love? Just tapped a Nook Horton. Fresh as the morning dew."

"That sounds perfect, Uncle Arthur," Bella said, leaning on the bar. "And don't let Tom off easy—he's been promising free drinks for as long as I can remember."

Arthur chuckled, sliding three pints across the bar. "Three as always? And might I say, you two roses ought to keep Andrew out of trouble tonight."

Andrew rolled his eyes, taking his pint with a wry smile. "You've been saying that for years, Arthur."

"And I'll keep saying it," Arthur replied with a grin. "Keeps you honest."

They made their way to their usual table by the fire, stopping along the way to exchange a few words with the locals. Bea paused to compliment Martha's knitted shawl, while Andrew accepted a lighthearted challenge to a game of darts from Old Harry. Bella, meanwhile, found herself cornered by young Tom, whose attempts at flirtation were as clumsy as ever.

"You know," Tom began, puffing out his chest, "if you ever tire of Bentley Hall, I've got a cozy cottage with a garden big enough for a Bentley rose."

Bella smirked. "Tom, your cottage is lovely, but I hear your roses are outnumbered by weeds."

The table burst into laughter as Tom turned red, scratching the back of his head. "Fair enough, Bella. Fair enough."

The trio finally settled into their seats, the firelight dancing across their faces. Bella sipped her pint, her gaze lingering on the globe. "You know," she said, "I think this is where I first realized the world was round."

Andrew raised an eyebrow. "Oh? Do tell."

Bella grinned. "The globe itself, the crash of a pint glass proving gravity, and, of course, physics—every action has an equal and opposite reaction, like when Uncle Arthur tosses someone out for mouthing off."

Bea burst into laughter. "A well-rounded education, I'd say."

Their laughter faded as Bella's thoughts turned inward. She remembered the times she had brought Payman to The Globe. The regulars had always been polite but cautious, and their subtle warnings now seemed to echo in her mind. "You know," she said quietly, "the only times I've ever felt out of place here were when I brought Payman. Did either of you notice anything?"

Bea and Andrew exchanged a glance. "It's not that they didn't like him," Bea began carefully. "It's more that they wanted to protect you. Some of them thought there was something… off about him."

Andrew nodded, his tone serious. "It's not just them, Bella. I've spent my life working with people all over the world. With Payman, I always felt like there was a piece missing. Like he wasn't being entirely honest. Payman just didn't present as consistently and straightforwardly as I would have liked. My intuition is there is something he's not telling us. You have met family members and I know you enjoy your time in London at his grandfather's home, but it raises questions." Bea and Andrew shared a long, concerned look. "Maybe it's time to reassess."

The words sunk deep into Bella's stomach. The past two months of silence had indeed weighed on her, no matter how much exploration, discovery, and relaxation she had within the estate, it had not distracted her from that nagging question about Payman's commitment to them, as a couple.

Bella stared into her glass, the weight of their words sinking in. Emotion overcame her and tears flowed. "You're right. I have been having my doubts and I haven't wanted to say anything. I don't know what to do. I haven't heard from him in 9 weeks since the accident. It's as if he has vanished."

Bea slid her chair over and hugged Bella. "It's okay. We're not saying anything conclusive. Only that we're just as concerned as you are. We're here for you."

Arthur saw what was happening and came over. He whispered to Andrew, "Is everything okay? Can I bring you anything?"

"I think a refill might be due already," Andrew said.

Arthur then asked Bella if she was okay. Bella wiped her eyes and took her time before answering. She smiled and said, "I'm hungry, and I'll bet you a pint that today's special is fresh-caught fish, chips, and mint peas. If I'm wrong, pull yourself a pint. If I'm right, bring three specials."

"Bella, you know you're every bit family here as your uncle is. Whatever you need, just ask."

"Thank you," Bella said gratefully.

"You're right anyway," Arthur said, sniffing in amusement. "Today's special is indeed fish. Three specials on the way!"

CHAPTER 21
THE MEHRI DYNASTY

Masters of Global Finance London, England
25 November 18:05 Local Time

Amir Mehri, the visionary patriarch of the Mehri dynasty, wielded financial influence that transcended borders. His sharp intellect and unparalleled acumen had built a sprawling international financial empire, enabling his family to rise among the ranks of the world's wealthiest elites. What set Amir apart was not merely his success but his ability to navigate the intricate labyrinth of global finance—sidestepping the sanctions that the international community had imposed on Iran. He constructed a complex network of offshore accounts, trust funds, and discreet financial entities. Billions of U.S. dollars flowed seamlessly into Swiss banks, Cayman Island accounts, and other secretive financial institutions.

This elaborate infrastructure of wealth preservation and growth allowed the Mehri family to operate with immunity from the geopolitical constraints that limited others in their position. Amir's genius lay not only in circumventing restrictions but in transforming challenges into opportunities, leveraging the family's influence to dominate global markets.

The Mehri family's wealth was more than a symbol of status—it was a tool of leverage. Their financial reach enabled them to invest in industries worldwide, from technology startups in Silicon Valley to luxury real estate in Paris, Dubai, and Hong Kong. Yet, Amir cultivated more than wealth; he cultivated power. His ability to sway markets, influence policy through strategic philanthropy, and foster connections with global leaders ensured that the Mehri name was not only recognized but revered.

Operating from his primary office address, 30 St. Mary Axe, popularly known as The Gherkin for its iconic architecture, Amir commanded his empire from one of London's most prestigious skyscrapers. The building, a beacon of financial power overlooking the city's elite districts, reflected the sophistication of Amir's operations. Within its sleek, state-of-the-art facilities, he brokered billion-dollar deals, cultivated relationships with powerful allies, and ensured the Mehri name remained synonymous with wealth, influence, and discretion.

Amir was not merely a financier; he was a master strategist harboring a deeply personal mission: the return of Iran to a secular government. Beneath his polished exterior as a global magnate lay a quiet determination to see his homeland shed theocracy for a future defined by freedom, progress, and democracy. This ambition, however, was shrouded in secrecy. Amir understood that openly opposing the regime would jeopardize the vast network of influence and alliances he had cultivated and built over decades. His primary ally in these clandestine endeavors was his grandson, Payman. Groomed from an early age, Payman was being prepared to assume leadership of the Mehri dynasty.

Amir surrounded himself with exceptional individuals, from his security personnel to his financial managers. His elite staff was a carefully curated group of individuals who reflected his own exacting standards and served as an extension of his power and influence. The team, composed of young, attractive men and women in their late twenties to early forties, was chosen for their peak physical fitness, mental sharpness, and adaptability. Each member embodied the values Amir prized most and had cultivated within himself.

Amir sought individuals with distinguished academic credentials—graduates from top universities in fields such as international relations, finance, engineering, and strategy. Their intelligence enabled them to anticipate and resolve complex challenges, whether in boardrooms or high-pressure situations.

He demanded aesthetic elegance. Each impeccably dressed in bespoke suits and designer attire. Amir's staff reflected the sophistication of their employer. Their polished appearances were tools for charm, negotiation, and influence. Their appearance opened doors to the world's most powerful individuals.

Recognizing the inherent risks of his operations, Amir ensured every team member was trained in martial arts, firearms, and tactical defense. They were equally adept at blending into social events or operating covertly in hostile territories.

Masters of subtlety, Amir's operatives navigated high-stakes environments with poise. Whether safeguarding assets, surveilling threats, or manipulating situations to achieve objectives, they excelled at their roles. Subterfuge and adaptability became their operational philosophy. They achieved the desired outcomes, by whatever means were required.

Payman, the heir to Amir's empire, was the model on which the team was molded. As Amir's successor, Payman held an invaluable seat at his grandfather's right hand, a position that guaranteed his influence in every facet of the dynasty.

Among Amir's closest advisors was Samira, his confidant, counselor, and chief strategist. A strikingly elegant woman in her early forties, Samira possessed a cunning intellect and sharp instincts. Fluent in five languages, she managed Amir's schedule and access, negotiated on his behalf, and ensured the seamless execution of his plans. Known for her icy composure and razor-sharp intellect, she was both a diplomat and a behind-the-scenes enforcer. Her natural magnetism and grace allowed her to wield her power in a man's world from a position of undeniable strength. Weapons were extensions of Samira's will—blade, rifle, it made no difference—but it was her finely honed body, swift and unrelenting, that delivered the final, most elegant strike. To Payman, she was a trusted mentor and an older sister figure.

One evening, as they sipped Persian coffee following dinner, at the Mehri compound in Kensington Palace Gardens, Samira broached a sensitive topic with Amir. It was not unusual for them to have difficult conversations, but this one required particular delicacy.

"Sir," she began, her voice steady yet cautious, "I must bring up a concern about Payman. I'm worried about the influence Bella has on him. I think he may be in love."

Amir, who had entertained similar suspicions, relied on his best leadership technique: asking questions and listening. "What gives you this impression, Samira?"

"You know I love him like a brother," she replied. "And as sisters often do, I've been watching him closely. He speaks of her constantly, there's a gleam in his eyes when he does, and I can feel his energy shifting. Bella has captivated him, and while she's important to us, I fear his judgment may be clouded."

"Tell me more," Amir urged. "Help me understand your intuition."

Samira leaned forward, her perceptive gaze unwavering. "It's simple. She's beautiful in a Spanish sort of way, his intellectual equal, challenging him and supporting him in his emotional growth. She clearly has dreams of a life with him, and I've seen this growing over time. With him rehabilitating at the family compound in Iran, I'm worried that isolation might be amplifying his attachment to Bella. I just don't want his emotions to distract him from his mission."

"I appreciate your insight. He needs to focus on his recovery and his mission," Amir said thoughtfully. "Being at the family compound in Iran should provide him with the clarity he needs. I trust your judgment. Guide him as you always do, but tread carefully. I don't want to disrupt our insight into the valuable work she's doing. Use your sisterly influence to keep him on track."

"That's a tall order, sir," she said with a faint smile. "Mitigating Cupid's influence is no small task. But I'll do my part. Perhaps being cooped up at the family compound in Iran has left him too much time to dwell on his feelings. A sailing weekend might be the perfect reset. He needs to get back at the helm of the 'Falkin after his accident."

"A brilliant idea," Amir agreed. "Arrange for a weekend outing, just the two of us. Thank you for your intuition."

With a plan in place, they turned their attention back to the day's other pressing business, their shared understanding deepening as they prepared to guide Payman through the challenges ahead.

CHAPTER 22
QUESTIONS

Bentley Hall Oxfordshire, England

25 November 19:40 Local Time

As they were leaving the Globe, Bea whispered in Bella's ear, "Are you up for an evening of 'girl talk' when we get home or should I just open a bottle of wine and let you talk?"

Bella giggled, "I wasn't aware girl talk was an ageless enjoyment."

"The wine is what makes it suitable for adults," Bea added without missing a beat. A dry tone cut through the loftiness of the invitation, "having it occur around a warm fire gives to the pretense that it's sensible. In any case, you don't think you're getting too old for it, do you?"

An hour later, the soft glow of firelight flickered across the room, casting long, dancing shadows on the walls. The fire crackled and popped in the hearth, releasing the occasional burst of warmth that wrapped around them like a comforting blanket. The faint scent of burning cedar filled the air, mingling with the fruity aroma of the wine in their glasses.

Bea sank back into her chair, the delicate creak of the leather faint under the ambient sounds of the room. Bella sat curled on the sofa, a wool blanket draped over her legs which were pulled up onto the comfy couch, her hands cradling her wineglass as if drawing heat from it. The soft clink of the glass as she set it down on the table punctuated the silence between them, a pause that felt neither empty nor awkward.

"It's perfect in here," Bella said softly, her voice barely above a whisper. She stared into the fire, watching the flames dance. "The kind of place where you can just be... let it all spill out."

Bea smiled, leaning forward to poke the fire with an iron rod. "That's what the fire is for," she said. "It listens without judgment. Just warmth and light, no matter what you bring to it."

Bella's lips twitched into a faint smile, but her eyes lingered on the flames. "I've given everything to him," she murmured. "Every part of me."

The warmth of the fire seemed to grow heavier in the room, wrapping itself around her words as she began to speak.

Bella spoke of how she had devoted herself wholeheartedly to their relationship, laying bare her soul to Payman and sharing every part of herself with him. Their days were sewn into a tapestry of shared experiences, moving seamlessly between the family estate and her cozy home on Magpie Lane. Together, they enjoyed holidays, worked side by side, and supported each other through the intricacies of academic life.

Bella confessed to Bea her desires and fears as she spoke of her life with Payman. Bea, sensing the vulnerability of the matter, made a point to ask Bella to breathe when her voice trembled.

"I'm sorry," Bella told her.

"You owe no apology," Bea said. "No one in life is prepared for their first love, worse still if it might well be the one."

Bella wiped her eyes. "It's a nice fragility to have. But I'm tired of the precariousness. I want stability."

Bea laid a hand upon Bella's arm and gave it a gentle squeeze. "We all go through it, we all long for the comfort to devote to another without fear. We long for the fragility because we want the feeling of being accepted for it."

At that, Bella wept a little more and Bea moved to her on the couch and put an arm around her.

"It's okay," she told her.

"I'm too buzzed for wisdom," Bella said.

They both fell apart laughing, Bella's face becoming a mess of tears that she then cleaned with her sleeve. After they embraced again, Bella explained how she devoted herself to their relationship and how Payman knew everything about her. She especially appreciated how Payman recognized her as an equal if not his intellectual superior—she found that many men in both her profession and in her routine encounters treated women as if they, men, were assigned superiority at birth. Payman brought each of these disrespectful types back into respectful order, although he knew Bella was more than capable of putting anyone in their place with her intellect and physically with her martial arts training that made her a lethal weapon. In conversations and faculty presentations, Payman frequently deferred to Bella's presence.

They met almost daily for tea or coffee at her favorite spot in Oxford, Blackwell's Books, directly across Broad Street from the watchful eyes of the Emperors and within sight of Sheldonian Theater. They smiled, laughed, cried and made love together. From Bella's perspective, she could envision her life with him. But could he make that same commitment?

She couldn't shake the feeling that something was amiss. He was notoriously secretive about his life before their meeting and the extended absences he spent in Iran. She understood that his presence was required in his homeland to leverage his education and research experience for the nation's benefit.

What dismayed Bella the most was his ghostly presence. Although he frequently contacted her online, professing that she was his top priority and expressing his desire to spend the rest of his life with her, he would then disappear for weeks without any communication, and whenever she offered to visit him in Iran, he presented a litany of excuses to deter her.

To cope with his absences, she threw herself into her own work, delving into the intricate and fundamental issues of subatomic particles and quickly became a central figure among the researchers, gaining a reputation within the inner circle of knowledge, innovation, and social activity. Many intelligent and attractive men vied for her attention, yet her heart remained steadfastly committed to Payman.

Bea, listened, commented and hugged Bella.

By the end of the evening, Bella felt a fragile yet undeniable clarity settle over her. The warmth of the fire, Bea's gentle words, and the unspoken weight of her own emotions had all worked together to untangle the knot in her chest. She knew now that the questions she had buried—out of fear, out of hope—could no longer be ignored.

As she stared into the dying embers of the fire, Bella recognized that the love she felt for Payman, as profound and consuming as it was, could not be sustained on fragments of his presence and promises alone. She longed for the life they had built in moments—moments that felt like forever—but she needed more than moments. She needed a partner who was fully present, fully hers.

Her heart ached with the weight of both love and uncertainty, but within that ache, she found resolve. It wasn't just about asking Payman if he could commit; it was about finding the courage to demand the life she deserved, even if it meant risking the one she had dreamed of.

Bella sat back in her chair, her wineglass empty in her hand, and let out a soft, steadying breath. Tomorrow, she would face the truth. And whatever came of it, she knew she couldn't keep walking the line between hope and doubt. Her path was clear now, even if the steps ahead were shrouded in shadows.

CHAPTER 23
MAGPIE LANE

Bella's Home Magpie Lane, High Street, Oxford, England 15 January 23:40 Local Time

Magpie Lane was a historic, narrow pedestrian lane in Central Oxford, running from High Street to Merton College. During the thirteenth century, it had been infamous for its prostitution, but in modern times, it had become classic Oxford—a highly desirable location centrally located near everything. It was a quaint lane that Oxonians loved and visitors couldn't miss on a tour.

Bella's narrow row house had been in the family for hundreds of years. She had played there as a child and stayed there on weekends while attending St. Rye's. Now, it was her home.

The cozy house on Magpie Lane seemed to breathe like a living thing, creaking and groaning as it settled into the night. Bella was asleep when a sound—soft but unnatural—stirred her awake. Her senses tightened as her eyes snapped open. She instinctively reached for the hunting knife she kept near her bed, its familiar hilt grounding her. It was the same knife her father had given her as an adolescent when he taught her independence in the natural world.

Quantum

The floorboards groaned faintly as footsteps approached, growing louder with each step. Her heart raced, and her grip on the knife tightened. She was certain it was an intruder. Then came the faint sound of a hand, maybe a ring, on the doorknob, followed by the soft twist of metal. The door creaked open. A man, she knew it was a man. No woman would dare enter another's domain. She readied the knife, knowing she would have the element of surprise. She steadied herself, her breathing silenced, ready to pounce with the agility and precision of a predator.

A shadowy figure entered the room. Bella's muscles tensed, coiled, every nerve ready to strike. But then, a scent drifted toward her, calming her tension like a soothing balm. Sandalwood, warm skin, and a familiar undertone of something uniquely his. Her grip on the knife slackened as relief and longing washed over her. She knew this man.

As he approached the bed, she reached for him. Her hand brushed against his skin, and she realized he was naked. Her touch moved instinctively, finding him, eliciting a low moan from his lips. Her body—once taut with fear—was now alight with anticipation. He had been gone too long.

She guided him closer, her hands exploring him with a mixture of urgency and tenderness. Her lips found him, and she moved with deliberate care, savoring the reunion. She knew every inch of him, every reaction she could draw from him, and she took her time, relishing the way his body responded to hers.

He, too, knew her well. His hands caressed her, their movements slow and reverent, as though rediscovering her. His fingers found their way to her thighs, smooth and inviting, and his touch ignited a fire within her. Their rhythm built naturally, guided by the intimacy of their shared knowledge. Every touch, every sound was a testament to the connection they had missed during their time apart.

As her body trembled with the waves of pleasure he drew from her, she reached for him, pulling him closer. Their movements became seamless, their passion unspoken yet deeply understood. They moved as if their bodies remembered the steps of a dance they had perfected long ago. Each shift, each motion, brought them closer to the culmination of their reunion.

When their shared crescendo finally came, it was accompanied by a profound sense of completeness. Spent and entangled, they collapsed onto the bed, their breaths mingling in the quiet aftermath. Bella nestled into his arms, her head resting against his chest, and let herself be enveloped by the warmth of the moment.

As sleep began to claim her, a single thought lingered in her mind: in his presence, the world felt whole again.

CHAPTER 24
JACK

University of Arizona Stadium UA Sands Club, community and faculty club
15 January 17:00 Local Time

Jack lived a life full of excitement. As a former U.S. Naval Communications Officer, he trained with the elite ranks of the U.S. Navy. Although his father was a hardcore U.S. Air Force officer who demanded excellence in everything Jack did and expected his son to follow him into the Air Force, Jack—ever the individualist—rebelled against his father's control and chose his own path in the U.S. Navy.

After graduating early from Arizona with his degree in computer science, Jack pursued his interest in exploration, believing the Navy was his window to the world. He served aboard the USS Benfold (DDG 65), an Arleigh Burke-class destroyer equipped with the AEGIS system, a critical defense against ballistic missile threats worldwide. These ships provide essential lines of defense in the Middle East, the Horn of Africa, the Indian Ocean, the Mediterranean and the Pacific. There was no shortage of adrenaline while being on standby for the launch of an Inter-

Continental Ballistic Missile. However, thankfully, these standbys for defensive launch intercepts are not the norm.

While serving on the Benfold, Jack became bored with the routine of shipboard life. Although most others would never call the routine on board a U.S. combat ship mundane, Jack was not a typical career naval officer. He craved adventure. Even the high stakes of preventing a rogue nation from launching Armageddon weren't enough to satisfy his thirst for adventure.

To satisfy his desire for excitement and danger, he volunteered to become part of the elite U.S. Navy VBSS service. VBSS stands for Visit, Board, Search, and Seizure. It was a critical capability of naval special warfare operations. It involved approaching and boarding a maritime vessel for the purpose of searching, and then possibly seizing contraband, peacefully engaging cooperative crew members, or effectively utilizing force to take down challengers. With many Navy SEALs assigned to the "War on Terror" working more on land, opportunities arose for shipboard personnel to conduct this potentially dangerous job of VBSS boarding and interdicting vessels at sea.

With this training, Jack transformed into an elite warfighter. His personal courage and rigorous training equipped him and his teammates with the skills needed to confidently board and interdict suspicious vessels on the open sea. This was truly an adrenaline-filled assignment. The approach of another ship on the open seas was always a harrowing experience. Day or night, rain or shine, sweltering equatorial ocean heat or the frigid waters of the North Seas, the team was ready for any assignment. Often, it required the interpersonal skills of a diplomat combined with the assertive style of a warfighter. These encounters ranged from refugee boats transporting people escaping tyranny or oppression, to fishing trawlers, to dangerous smugglers and spies. The team engaged with both cooperative and uncooperative subjects, handling each situation with precision and expertise. Jack's personal experiences were diverse, ranging from humanitarian missions that provided aid to those in need, to high-stakes operations that involved taking down dangerous adversaries.

Upon completing his service commitment, Jack utilized his GI Bill benefits to further his education, focusing on his burgeoning interest in nanotechnology and material sciences. He pursued and completed a PhD from the University of Arizona, where his cutting-edge research quickly garnered attention. Recognizing his potential and expertise, his alma mater promptly hired him as a prominent research scientist, allowing him to continue pushing the boundaries of scientific innovation.

That evening, Jack eagerly awaited his new mentor, Dr. Stephens, a distinguished expert in the field of quantum networking and computing who had recently joined the team as a valuable national asset of knowledge. The two were set to convene for a dinner meeting at the University of Arizona Faculty & Community Club.

The club was situated within the stadium club facility overlooking the grand Arizona football stadium. From this vantage point, they could enjoy sweeping northern views of the picturesque UofA campus and the majestic Catalina Mountain range, which stood prominently in the background. These mountains were often described as embodying the essence of "purple mountains majesties," especially when illuminated by the vibrant purple hues that adorned them during the breathtaking Tucson, Sonora Desert sunsets, reminiscent of the imagery evoked in the national anthem, America The Beautiful. The sky was now fully ablaze in a painter's palette of reds, oranges, blues, and violets. It was an amazing sunset, with the mountains bathed in pervasive purple majesty. Bathed in the vivid glow of the desert sunsets, the mountains and foothills formed a scene as stirring and timeless as the anthem itself—a daily reminder of the beauty that inspired a nation's song.

To the south, the club offered a view of the expansive UofA football stadium with a seating capacity of fifty thousand, where many memorable UofA football battles had taken place. Even now, in the offseason, players were on the field warming up, practicing field goal kicks and downfield punts. The energy and excitement of the club were palpable.

Nearby tables were occupied by former NFL stars who were now part of the UofA football staff, as well as alumni and supporters of the university. Jack seamlessly blended in with the esteemed faculty members of UofA. These esteemed members of the university community and influential members of the public frequented the club for meals or meetings to engage in discussions of local and national importance. Many of these individuals were actively involved in initiatives aimed at translating UofA's cutting-edge research into innovative startup ventures that had the potential to generate substantial economic growth and opportunities.

Looking to his wrist to check the time, Jack couldn't help but beam with pride every time he glanced at his Rolex Milgauss, which his father had presented to him upon his graduation with his PhD from UA in material science associated with quantum computing. The Rolex Milgauss was developed in the 1950s and released in 1956 for the scientific community, doctors, engineers, and scientists working in laboratory environments who might be subject to electromagnetic fields. These fields would disrupt any other mechanical watches. Its name is a combination of the French word Mille for thousand and Gauss for magnetic induction.

Magnetic fields are everywhere, and this watch ensured that it would keep Jack on time. Jack especially liked the lightning bolt second hand sweeping the back dial under the green crystal. The crystal was one of Rolex's many characteristics that made them stand out in such subtle ways.

The watch was certified by CERN, the Conseil européen pour la Recherche nucléaire (European Council for Nuclear Research). The same CERN that oversaw the world's largest and most advanced particle accelerator, the Large Hadron Collider. What made this watch resistant to electromagnetics was that the watch had a soft iron shield or Faraday cage surrounding the nonmagnetic alloy movement. It was unlike any other watch, the only Rolex to feature a unique green sapphire crystal. Between its inner workings, the unique visible green crystal, and the orange lightning bolt second hand, the Milgauss was a real eye-catcher. Jack beamed every time he checked the time.

Just then, Jack looked up from admiring his watch to see his new boss, Dr. Stephens, VP of Research and Development, Quantum Computing. Dr. Stephens had stopped at a table of National Football League stars, shaking hands around. This was a good sign for Jack, as Dr. Stephens looked affable and friendly. His new boss was an extrovert and not afraid to engage with the club celebrities. Dr. Stephens made eye contact and invited Jack over to the table.

Introductions were made around. At the table was a gathering of four famous NFL stars, all alumni of UA: Rob Gronkowski, with 11 seasons in the NFL, 4 Super Bowl Rings, and the holder of the world record for "catching a pass, a dropped football, from a helicopter 600 feet in the air" achieved in the very UofA football stadium now visible to the south of the club; Nick

Foles, quarterback and Super Bowl LII MVP, who led the Eagles to a victory over the Patriots; Ricky Hunley, Pro Bowl defensive lineman and feared pass rusher; along with Chuck Cecil, Pro Bowler and reputedly the toughest defensive back to ever play the game. Jack was in awe of the talented players he was meeting. Although Jack may have been seen by some as a scientific nerd, he was a man's man and knew his football. He made it a point to understand the rich history of UofA football as a student.

Dr. Stephens introduced Jack to the group, saying, "Jack is our own Pro Bowl contender to change the world of computing as we know it. He, like all of you, is a UofA alum. He's also a Navy vet, so show him some respect."

All four NFL Pro Bowlers stood at once to thank Jack for his service to the country. Ricky spoke first, "You've accomplished something none of us can brag about. You're keeping the world safe from terrorists and letting guys like us make a living playing football, the game we love."

All agreed. Jack blushed at the admiration. They all laughed, and Jack shook hands with each. "We warfighters always appreciate watching you guys play a game we love, showing finesse on the field and making amazing plays. And you get paid to play!"

Gronkowski asked Jack about his favorite Super Bowl. Jack replied, "I'm not sure it would be wise to choose a favorite with you guys all standing taller and stouter than me."

Cecil replied, "I think you're taller than me and bigger too, even in my playing days."

Everyone laughed. Ricky added, "Heart counts too in bigness, Cecil, and you've got that covered."

Gronkowski chimed back in, "You can't get away without answering the Super Bowl question."

Jack caught his breath, not only from being in front of these four famous NFL stars, but also in preparation for telling this story. He continued, "I'd rather tell you about my worst Super Bowl, LII, 2018, the one you pulled out, Nick. We were on deployment in February off the coast of Somalia with decent weather and calm seas around the Horn of Africa. The Commanding Officer was a huge Eagles fan. He had the entire ship pumped up for the game. You know there was a lot of chiding and ribbing about the game. The crew loved the banter. Also, this was the only time you could chide and banter with the skipper."

The guys laughed and poked some joking at Foles. Jack continued, "Days before the Super Bowl, our Super High Frequency (SHF) antenna system went out because of an amplifier, and we lost the Direct TV signal. I know that's a lot of geeky technical stuff, but bottom line, we lost our ability to watch the game. To watch you two go head to head. Unfortunately, the amplifier replacement was unavailable. Electronic technicians rushed to repair the amplifier. They climbed the main mast to reach the amplifier. But once it was in the shop, we determined there was no way we could fix it.

"So the CO, the skipper, pulled strings and made a parts swap from another ship with a spare amp. The amp was helicoptered out to our ship. The helo crew dropped the part off on the fantail flight deck. I was never happier to see a helo arrive. That was a time I really appreciated the efficiency of the Navy resupply mission. Hours before the game, we were working to bring the Direct TV back up as the crew became more and more agitated in anticipation. We got the signal back, and minutes before the game, the picture went live. We were the heroes of the day. The crew cheered, ate like it was a Super Bowl party—but no beers. They all loved it, and the CO went to bed that night happy with a win for the Eagles."

CHAPTER 25
JACK & DR. STEPHENS

University of Arizona Stadium UA Sands Club,
community and faculty club
15 January 17:30 Local Time

The two of them settled into dinner and drinks while enjoying the atmosphere of the club. They engaged in laughter, shared stories of their upbringing and life experiences, and avoided technical talk as they began to get to know one another. Jack had been with the university for his graduate studies, his post-graduate work, and now held a full-time faculty appointment. Dr. Stephens was relatively new to the university and eager to get to know both Jack and his expertise beyond the university and campus.

"Jack, I'm very pleased to get to know you. I hear good things about you and your work. I understand from the team that you're exploring some groundbreaking projects. Also, please don't call me Dr. Stephens; call me Rick. Let's be on a first-name basis as we have a lot of 'roll-up-our-sleeves' work to do. Long hours in the lab and in discussions over a wine or bourbon before we reach our quantum networking stability objective. I want to be considered a member of your team."

Jack smiled. "Thanks, Rick. I appreciate your comments and way of thinking. As you know, it's a team effort. Not just me. Yes, we are doing some groundbreaking work in exploring new crystalline materials that form the very basis of quantum, both computing and networking. As you know, it's going to take a breakthrough to unlock the possibility of developing a replacement for silicon-based conductors."

The two sipped Woodford Manhattans, enjoying one another's company and the atmosphere. Rick continued, "Ever since the UA received the honor of being named a member of the National Information Science Research Center, becoming a grantee of the funds to provide the new Superconducting Quantum Materials and Systems, I've known UA is where I wanted to work to make the material breakthrough. We have the talent, the labs, and the enthusiasm. I knew the breakthrough would be at the atomic level or the nano level, and that's where we'll make a difference. We'll overcome the limiting mechanisms of qubits."

"OK, now you're getting sexy with that kind of talk," replied Jack. They both laughed. Jack continued, "So I've read your bio, but who is the real Dr. Stephens, Rick?"

"I had been doing this work for probably longer than you had been on the planet. I started with Bell Labs when it was the premier high-tech research organization in the world. I'm not talking about Mr. Bell's invention, the telephone, but things like the transistor, probably the most important invention of the 20th century; contributing to lasers, communications, satellites, missiles, and such a practical application as solar cells. I'd like to take credit for those, but they were all before my time!"

"Haha, I know you've got Nobel Prize-worthy knowledge, but no way I'd ever accept that you were around for the invention of any of those. You're still a young guy. Anyone in this field has to be young, especially from a mental standpoint," Jack replied.

Rick continued, "I've also been a venture capitalist, engaging with technologists responsible for some very interesting breakthroughs. I've worked with the Department of Defense and some of our letter agencies in advancing technologies. Just as with the breakthroughs at Bell, our work will lead to defense and industrial applications, and we're targets for espionage. Nations, interest groups, and industrial competition all want to make traction in quantum. We must keep our knowledge safe. We have something else in common; we've both fought the war on terrorism."

Rick added, "Jack, that was an interesting story about the Super Bowl and shipboard life. I understand you were also a special forces guy."

Always modest, Jack commented, "Not special forces that we think about like SEALS or Delta Force, but Navy VBSS—Visit, Board, Search, and Seizure. With so many SEALS on land doing joint operations, it's become much more in demand. Kind of a volunteer special operation. We were highly trained and equipped with the best gear available. It's normally about gathering local intelligence through contacting crews of commercial fishing or cargo ships that traverse the waterways. Highly publicized waters such as the Straits of Malacca between Malaysia and Singapore or off the Horn of Africa between Yemen and Somalia. The Somali pirates are probably the best known to us in the US. There are really two types of 21st-century pirates: small-time bands looking to capture cargo and revenge attacks, interested in quick hits and quick loot, similar to the old pirates of the past. Then there are the organizations of pirates—those who syndicate their operations so that a community or group contributes cash to the operation and then shares in the bounty. Kind of like locals investing in a business operation and sharing in the success. These groups often go after the bigger vessels to increase potential return and hold crews for ransom. Then, of course, all around the world are human traffickers. We had experiences with all of these operations."

"Wow, you know your stuff. And yes, I see the difference. I remember the infamous Captain Phillips hijacking," said Rick as he urged Jack to continue.

Just then, the NFL players got up to leave but stopped by the table to say their goodbyes, again thanking Jack for his service.

Rick replied, "You guys need to invite Jack to lunch and have him tell you some of his pirate hunting stories."

Ricky responded for the crew, "We'd love to, none of us like the Buccaneers!"

A laugh went around, and the guys left. The club was clearing out, leaving only those engrossed in dinners or reading or working on their computers sprinkled around the room. Rick urged Jack to continue, "Go on, you were just going to talk about some of your experiences."

"First, we were the 'tip of the spear.' We were patrolling the Gulf of Aden, between Yemen and Somalia. Second, the pirates ran the gamut from local bad guys with AK47s and fast boats to highly organized and well-funded cartels. You've heard about the Somali pirates. They were known for daily threats to shipping. You remember Captain Phillips and the movie where he and his ship, the tanker Alabama, were overcome by well-organized Somali pirates. We were the first ship on scene as our return trip from deployment was a couple of days away."

"Yeah, I remember that movie. Did you go in and attempt a rescue?"

"No, there were negotiations ongoing, so we were assigned an observation role. Several days later, Captain Phillips was transferred to one of the ship's lifeboats, and it ended up in tow behind one of our sister ships, the Bainbridge. Later that day, it was believed that Captain Phillips was in immediate danger, and US Navy SEAL snipers acted. When all three pirates were visible, the snipers took them out, leaving Captain Phillips unharmed. The rest is history. Quite an experience for both sides."

"So does your experience include sniper training?"

"No, we, of course, had to be proficient and certified in all weapons, from rifles to the use of our hands. But our most important training was that of compassion. We had to remember we were dealing with few pirates and mostly with people trying to survive life."

Yes, I would imagine that survival is difficult in that part of the world, from warlords, gangsters, pirates to famine and drought."

"Probably the most memorable VBSS experience was actually a humanitarian missions. We performed many, many humanitarian missions. We were doing routine patrols. We saw a small skiff on the horizon. That was unusual for a small vessel to be in a shipping lane. As we approached, it became clear that the boat was dangerously overcrowded with people. Cautiously, we approached because we weren't sure if it was a setup drawing us in only to ambush us. The bad guys have a strategy of drawing us in and then trying to take us out or inflict casualties on the Americans using human hostages. As we closed in, there were dozens of people in the small space, covering every inch of the skiff, including under the seats and under the deck. We always traveled with a local civilian translator. These were trusted, caring locals who were our ears and voice in otherwise unknown languages or dialects. We neared the craft. The people were waving, cheering, and welcoming us. We tied up and stabilized the overfilled craft. Then our translator informed us that they fought off the bad guys who were taking them to sell on a nearby island. The bad guys were now shark food, dumped into the ocean. We provided the water and food we had on board and called for additional medical, food, water, and shade cloths. Our ship ferried survival aid, and we contacted the local coast guard to recover this group of humans, abandoned on the open sea to die or eke out survival as they navigated to an unknown future on an unfriendly coast. We stayed with them, making friends, providing flashlights, duct tape, and candy to the kids. This was a success. We saved lives."

As Jack ended his story, it was clear the humanitarian effort had made an impact as tears welled up in the corner of his eyes. He smiled with pride as he discreetly blotted the tears.

CHAPTER 26
WHAT'S NEXT

Oxford University, Oxford England

12 - 13 March

Work, research, department meetings, tutoring eager young students and long nights in the physics labs took their toll on Bella and Payman. They had fallen into a routine of working, not seeing one another all week, then weekend coffees at Blackwell's books, each browsing their own areas of interest, meeting back with their selections for a coffee or tea and quietly reading for an hour or so. Saturday and Sunday evenings were spent on nice dinners or drinks with colleagues. The sex had become just like their relationship, unimaginative but comfortable, and then it was back to the academic's and researcher's lives for another week.

Payman grew distant and preoccupied. Bella questioned what had happened to their excitement and the man who was so taken by her. She was never really comfortable with his past, his family and why she had not been invited to visit his home or the opportunity to get to know his family, though she had met his grandfather in London. His billions didn't make him distant but rather he was inclusive and offered opportunities for visits and dinners even when Payman was back in Iran. But when Payman was with his grandfather, he became different quiet and withdrawn.

Bella, reflecting on her heart-to-heart conversation with Bea, took him to the Turf Tavern where their love story had commenced years prior. Seated across from each other with pints in hand, she composed herself.

"You know that I love you, and I've never encountered someone who challenges and supports me in the way you do," Bella began. "But I must admit that we have fallen into a predictable routine. We tend to focus solely on our work, sharing little about our successes, failures, and research. I learn about your projects from your colleagues. While I adore your grandfather, I realize I know very little about your family or your home. Where do you think we stand, Payman? What direction are we heading in?"

Payman met her eyes and took his time in answering. "I understand what you're feeling. I share the same sentiments, and I can't say that I haven't been worried myself. I cherish you and fear losing you. Lately, I've been preoccupied and under pressure from my sponsors to deliver."

Bella reached out for his hands across the table, intertwining their fingers. Sensing the unusual warmth in his hands, she wondered if her initiating the conversation had triggered something under his surface. She asked him, "What's troubling you?"

Payman briefly distracted himself, looking away from Bella and watching the passers-by around them. His face sank, an exhausted frown descending upon him. "I'm anxious," he said. "All the time, I feel anxious. I fear losing you and I can't bear the thought of that."

"You're worry is correct, either we grow together or you are going to lose me," Bella insisted. "I want to know more about you. Be more involved in your life. I want to be one with you. If there's anything holding you back or anything about that that worries you, I'm here to listen."

"I've been distracted, but I appreciate your belief in us," Payman expressed sincerely.

"Payman, what do you mean? " Bella pressed for further clarity.

As Payman spent a long moment with his thoughts, Bella was concerned that he would remain reserved and this conversation would take them nowhere. He did, however, give her an assuring wide smile.

"Maybe it's Oxford, It's dreary, chilly, and dim here. Let's escape and head towards the sun." He pinched his tongue between his teeth while observing Bella excitingly raise an eyebrow. Then he proposed, "How about we go to Crans-Montana? I know how much you enjoy skiing, and Crans-Montana is one of the sunniest destinations in the Swiss Alps. You'll adore it. We have a family chateau right on the slopes."

Bella, if anything relieved that he'd made some sort of advancement with the topic, responded,

"Yes. Let's be somewhere where we can get away from the pressures of daily life, focus on us, and enjoy the mountains and the sun."

Beneath the persistent shroud of gray winter skies, the two souls, united by what Bella believed to be love and a mutual passion for one another and for physics, made a decision to trade the academic environment of Oxford University for the tranquility of the Alps.

Two days later their journey commenced on a damp, gray morning, the air cool and thick with moisture as they embarked on a train journey bound for London Heathrow. The rhythmic harmony of the train mirrored the excited palpitations of their hearts. Embracing the experience and the connection it could bring, Bella remained open-minded and held optimistic hopes that this trip would be the rejuvenation they so sorely needed.

The transition from the train to the luxurious comforts of first-class air travel marked the true inception of their escape. Welcomed with warm smiles and a flute of champagne, they settled into their seats, enveloped in a cocoon of opulence amidst the clouds. The flight itself became an adventure—a gourmet meal and a selection of wines that hinted at distant vineyards. In-flight entertainment and plush amenities served as a backdrop to their shared laughter and eager discussions about the slopes that awaited them.

Upon their arrival, they were whisked away from the bustling airport by a waiting black BMW SUV. The ride from the airport, took them to the serene elegance of Crans-Montana, a gem nestled in the heart of the Alps. Each twist and turn of the journey unveiling majestic granite peaks shimmering with crystalline textures, snow-draped valleys stretching endlessly under a canopy of clear blue skies, and serene forests cloaked in pristine winter white. Here and there, the stillness of the landscape was brought to life by fleeting glimpses of wildlife, adding a touch of enchantment to the already captivating panorama. The slope-side chalets emerged as a breathtaking wonder.

When they arrived at their accommodation, Bella was greeted by the crisp, clean aroma of pine trees and the feel of snow dust on her skin. Fully allowing herself to be here, she appreciated Payman's intuition about the need to escape for a while.

Their chalet, a sprawling sanctuary of warmth and rustic luxury, was perched on the side of a slope, ski in, ski out accommodations. The interior harmonized traditional Alpine aesthetics with modern conveniences and a crackling fire in the fireplace at its center.

Bella, on the patio, was lost in the mesmerizing beauty as the winter wonderland unfolded before her. Payman brought her a generous goblet of red wine. "You seem to complement the breathtaking view perfectly," he said.

"You haven't wasted any time," Bella said. She took a sip of the wine and bit her lip as Payman held her by the waist and kissed her. Emotions flared in her, she felt Payman reciprocate.

"Livio Sassetti Brunello di Montalcino Pertimali from Tuscany, the wine that is…" he explained. "a personal favorite introduced to me by Rob while sailing on the 'Falkin.'"

"One of your many up-standing acquaintances." The aroma of the wine made Bella crave just a little more. She turned to the view through tall chalet patio windows. They observed the final skiers gracefully descending the smooth slope back to the lodge and the alpenglow cast by the setting sun high in the mountains, illuminating the granite cliffs with a rosy hue. The trees glowed as the sunlight danced off the snow beneath them. Bella's attention was drawn to a pair of snowmobiles stationed along the tree line between their chalet and the neighboring one up the mountain.

"Payman, look at those guys on the snowmobiles. I remember seeing them at the airport, and now they are near our chalet," Bella remarked.

"Don't fret, they are part of my grandfather's security team. Whenever we travel, the security detail is always close by. They normally excel at blending in and remaining inconspicuous. Rest assured, there is no cause for concern. Tonight marks the beginning of our time together. Let's savor a delightful dinner. I've arranged for a chef to prepare a scrumptious mountain meal. I know you enjoy spaghetti carbonara with lobster bisque to start, followed by tiramisu for dessert. We can then unwind by the fire with a glass of port," Payman reassured her.

Bella replied with just a bit of seduction in her voice.

"Wonderful, I look forward to the meal but can't wait for the fireplace."

Dinner was amazing from the bisque that tasted like a spoonful of succulent lobster in each bite. Bella who was a great cook in her own right knew that the chef had reduced the stock over days, not minutes. Wines were paired with attention to what would complement each dish. The dinner seemed to top off a wonderful day together.

Settling into the warmth of the fireplace following dinner. She marveled at the natural stonework of the massive fireplace. The fireplace was river rock three stories tall directly to the beamed ceiling, with two balconies surrounding the great room. The windows were a view unto the night world of the mountain, the cable-trams were still running taking diners to the multi-star bistros above. The snow cats were grooming the slopes. The lights twinkled across the horizon and the granite cliffs were now reflecting the moonlight. It felt like a fairy tale land in her own snow globe.

Her thoughts returned to the soon to be delivered tray of after dinner delights. The fire crackled, the cedar scent of the wood from the fire was in the room. The leather, wood, rock and pleasant smoke was an aphrodisiac for her. She loved the scent of room, she was sure her pheromones were sending a similar signal to that which she was receiving from the atmosphere and her anticipation of Payman's return. She quickly took several thick plush throws, maybe merino wool or cashmere, blankets off of the couch and laid them on the floor in front of the fireplace. It was clear Bella had thoughts other than food on her mind. She created a little nest in the warmth of the fire.

She settled into the luxury of soft blankets. The combined thickness of the plush blankets were a giant pillow on the worn wood floor. The nearby coffee table would be the perfect spot for the tray. But her thoughts were far beyond the food. She readied herself, removed her bra without disturbing her blouse, removed her socks and let her shiny raven hair fall around her shoulders. Her perfume subtle. Who could resist such an alluring woman.

As Payman turned the corner into the great room, his senses were immediately heightened. He knew the tiramisu could wait. Setting the tray on the table, he joined Bella in the nest she had created near the warmth of the fire. She greeted him with a passionate kiss, one that reignited the spark between them and hinted at the evening ahead.

He sank into the softness of the blankets as Bella's hands explored his chest with deliberate tenderness. Her movements were slow and measured. She was glad he wore a buttoned shirt. Slowly she unbuttoned every one then equally slowly unzipped his pants. She sensually caressed every muscle in his chest and taunt stomach. She let her fingers trail over his defined muscles, admiring the strength of his athletic form. He had the body of an athlete, muscles rippled, he did not possess the body of someone who spent his life in a laboratory. Payman, captivated by her confidence and grace, allowed himself to surrender to the moment. The crackling fire mirrored the heat building between them.

He wanted to kiss her but she resisted, in complete control. He was under her spell. On his bared chest she gently then more aggressively licked and softly bit his neck and chest. He was tingly, how could these simple movements arouse him to the point of explosion and she was only at his chest. Payman wasn't sure if it was the heat of Bella or the fire but he grew warmer, beads of sweat formed at his forehead. Bella noticed and licked them off.

"I love your taste, I love your feral aroma when you get excited and I love your manhood in my vagina. But you'll have to wait for that." Bella almost purred as she gently touched his lip, not wanting him to reply.

Bella's allure was undeniable. She swayed with a quiet elegance, removing the last pieces of her clothing in a way that felt both natural and deliberate. The firelight danced across her skin, accentuating her curves and beauty. Payman reached for her, but she playfully held him back, taking her time and savoring every moment.

When she leaned forward, her hair brushing against his chest, he felt a shiver of anticipation. Their connection deepened as their hands and lips explored one another, guided by the quiet intimacy of the moment. The rhythm of their movements was unhurried, a dance of trust and desire that built steadily like the warmth of the glowing embers in the fireplace.

The rhythms of their motions were natural and complimentary. Their passion built steadily as they finally gave in to one another, their bodies and hearts moving in perfect harmony. Then the crescendo they were rocked with quivering delight, that seemed endless.

The world outside faded away, leaving only the crackle of the fire and the deep connection they shared. When the moment passed, they lay together, wrapped in the cashmere blankets, breathing in sync as the fire's glow softened around them.

Dessert went untouched as they rested and exchanged gentle kisses, their love reaffirmed in the quiet aftermath. They eventually fell asleep in each other's arms, the warmth of the room wrapping around them like a cocoon.

When Payman awoke, the embers in the fireplace cast a soft light across the room. He stood and approached the large picture window, gazing out at the enchanting sight of snowflakes drifting silently from the dark sky. Each flake seemed to dance on invisible currents, that created a mesmerizing display.

Payman knew the fresh snowfall promised a perfect day on the slopes. Smiling at the thought, he returned to Bella, gently lifting her in his arms and carrying her to bed. She wrapped herself around him, her trust and love evident in her relaxed embrace. As he settled her under the covers, he felt a quiet joy. They slept soundly, their dreams filled with love and the promise of new adventures.

CHAPTER 27
MOUNTAIN GLORY

Crans-Montana Swiss Alps, Switzerland
14 March 07:42 Local Time

Cocooned in the fluffiest down comforter each wrapped in the other's arms, muffled explosions woke them. Bella bolted herself upright. Payman gently placed his hand on her lap and stroked her lovingly.

"Avalanche control," he told her. "It snowed last night. I'll bet it dumped a ton of new powder." He stretched his back and yawned widely. It's going to be an epic day on the mountain."

Bella pulled the curtains open, and the light of the alpine day poured in. Snowflakes danced across the window, glinting in the sun's rays. Light, powdery snow the kind that whispered under skis and swirled in the air at the slightest disturbance, forming a translucent veil across the sky and transforming the rugged alpine terrain into a dreamscape.

"It's beautiful," Bella said standing against the window, a soft, still expression on her face giving away her inner serenity. "Do you think people use it as their happy place? You know, for when times are tough?"

"Without a doubt," Payman said. "I don't come here enough. I love it. Do you?"

Bella turned to him, her morning robe slipping from her shoulders. "I love you more. Come and find out just how much."

Payman couldn't resist her invitation. She slipped back into the cocoon of blankets, their bodies immediately warming, their breath quickening with anticipation. The day could wait.

The day continued with their journeying through the crisp morning air. It bit at their cheeks as they boarded the tram and ascended through the breathtaking canvas of snow-draped mountains under the glistening sun. They joined the excited crowd of skiers all eager to carve their first tracks. As the tram climbed higher, they soaked in the serene stillness punctuated only by the distant calls of alpine birds and the soft hum and clicks of the cable.

As they stepped off the tram at the summit, they looked in awe and with anticipation at the expanse of pristine slopes as the sun cast diamond-like sparkles across the snow, and their anticipation was rewarded with their first run. Bella felt exhilarated. The snow crunched under her skis, every turn a graceful dance between her and the mountain.

On the groomed runs, her perfectly tuned skis sang a melody only seasoned skiers know—the soft whistle of skill and precision. The sound that perfectly tuned skis, ridden by an excellent skier make. A sound that only those skiers who have perfected the art have the joy of hearing. Bella listened to the song of her skis on the groomed soft snow. The fluid view of nature around her momentarily cleared her mind of worry as she focused only on the heart-racing descent. She was a bird in flight soaring above the valley that stretched out before her. She loved the long runs, the thrill of the descent and the beauty of the wilderness. She lost herself in a moment of pure bliss, a culmination of anticipation and desire.

The mountains called to them, and they answered with courage and exhilaration. Together, they spent the day gliding down slopes, their laughter and whoops carried away by the wind. Payman led Bella through open trees where the powder was untouched. The wide spacing allowed for cautious speed, but he remained mindful to stay focused, avoiding distractions.

The dark trunks of the trees became their guides. Payman glided on the powder in a line carving between the trees. The more he focused on avoiding the trees the more he became caught in the trees controlling his line and the less he was in control. Payman became caught in a line between several trees. The more he tried to avoid them, the less control he had over his path. Attempting a clean turn, his ski edge caught, and he tumbled. His skis popped off, and his helmeted head disappeared into the snow.

Bella heard his muffled yell as he plunged headfirst. She expected the sight of a "yard sale," with equipment scattered across the slope, but the heavy powder had swallowed everything, including Payman.

His tracks disappeared fifty meters ahead. Carefully, Bella slowed herself and approached the spot where he'd vanished. She shouted, "Payman, are you okay?"

A muffled voice replied, "I'm here. I'm okay!"

"I can't see you! I can't see any sight of you." she called.

"I'm in a tree well! Can you follow my voice?"

Tree wells, areas of deep, soft snow surrounding trees, can be dangerous. Being caught in a tree well can be frightening. It's the area surrounding a tree and the well is created by the snow blowing around the tree which leaves a vacant area around the tree trunk The well can be many feet deep with walls that could easily come crushing down.

Bella removed her skis and placed them tail-first in an X in the deep snow. Using her poles for balance, she hiked through waist-deep powder. Step by step, she worked her way closer. It was a slow go. A step, sink in, balance, then all over again. Eventually, she saw his blue helmet rising near the tree just below snow height. Only by his helmet was he visible. His body buried in the powdery well.

"Payman, I see you. I see your helmet that is. Stay still," she instructed.

Bella removed her pack and pulled out a yellow climbing rope. Her father had taught her mountain safety—from winter survival to climbing to off-piste skiing, she was prepared. She found the yellow climbing rope. She secured one end to a nearby tree.

"Look up!" she said. "Throw your poles and skis up first, then use the rope to climb out."

"Got it," Payman replied. "Watch out—here they come!" He threw his poles and skis, which Bella set aside. Struggling against the soft snow, Payman used the rope to pull himself up. It was slow and exhausting, but he finally emerged, collapsing on his back in the powder, panting for breath.

"I owe you dinner," he said with a chuckle, brushing snow off his neck and gathering his gear. Breathing deeply to catch his breath.

Bella smirked as she coiled the rope and stowed it in her pack. "Dinner is a given. You owe me a massage."

Together they traversed the mountain to a mountainside cantilevered bistro, La Roca, with its amazing views and its traditional dishes. It was a favorite among those who knew the mountain. They sat outside on the patio, bathed in sunlight, and drank in the spectacular vantage point over the slopes. They indulge in a gourmet meal that married local flavors with international finesse, each dish a testament to the chef's artistry. For Payman, each course was paired with a glass of fine local wine. Lunch became not just a meal but an experience, one that delighted his senses and complemented the day's adventures.

"Are you sure you're okay," Bella asked noticing he was awkwardly holding his knife and seemed to struggle to cut through his food.

"It's a sprain, I think," Payman said wincing. "I'll need to rest it." Before Bella could show sympathy, he insisted, "Don't let it get in the way. You should still enjoy the slopes.

Bella took his hand and kissed it. "Rest well then. If I have any energy left tonight, you're putting it to good use."

Bella sped off to her own adventure. She felt like she was in heaven. The challenge of the runs was surpassed only by their reward—a rare, intimate connection with nature. It had been a while since she had skied and doing so in conditions like these were rare and precious, a deep connection with nature that was both intimate and grand. The deep snow demanded boldness, pushing her limits and sharpening her focus. She stayed on patrolled runs, her skis carving through the pristine powder with effortless grace. She was 'in-the-zone' where only the mountain in its glory and Bella existed in this time and place. Her body was on autopilot at speeds unknown to most skiers. Her legs burned as she maintained her balance, but the endorphins flooded her body with a unique sense of peace.

The sun dipped behind the rugged peaks, casting an alpenglow across the slopes.

Bella wound the day down with a final run. The snow had turned to a sparkling gold in the fading light. On a wide-open run, Bella straight-lined the fall line. She was a rocket. She imagined that this might be how an eagle must feel as she monitors the world below. She continued faster and faster, only slowing as she passed a family group taking their time on a slow cruise. She opened up again with the fall line pulling her below. This was her happy place, enveloped in a deep sense of contentment, the day's memories etched into her heart. She remembered where she was headed and traversed the slope to her right toward the chalet. She held her speed until the last moment then came to an almost immediate hockey stop, holding her balance as she became almost parallel with the slope, creating a vast cloud of snow spray.

She smiled, loving the day, basking in her endorphins and the beauty of nature.

CHAPTER 28
THE AFTERGLOW

Crans-Montana Swiss Alps, Switzerland
14 March 18:00 Local Time

Payman, much to Bella's delight, arranged a couples massage at the spa. After a full day of carving through the pristine slopes of the Alps, Bella's muscles ached but her spirits soared from the exhilaration of the descent and the beauty of the snow-capped peaks. She savored the thought of a relaxing hour on the massage table. The air was thick with the soothing scents of lavender and eucalyptus, signaling it was time to unwind.

They were guided into a tranquil room bathed in soft, amber light, that enveloped them in warmth. The ambient sounds of nature—a gentle stream, the soft whisper of wind through trees—filled the space. Bella and Payman exchanged smiles, a silent acknowledgment of the shared experience to come. They settled onto the tables and let the skilled therapists begin their work. With hands warmed by oils infused with alpine herbs, the therapists began with gentle strokes, easing into the layers of muscle fatigued by the day's adventures.

The therapists' techniques were deliberate, targeting areas burdened by the physical demands of skiing—shoulders drawn tight from the focus on holding weight forward over the skis and the grip of the poles. Legs pulsing from the constant push and pull of the skis against the snow. Calves screaming with built-up lactic acid. The therapists' synchronized movements allowed the couple to drift closer to a state of shared bliss. Knots and strains dissolved under the experts' touch, replaced by weightlessness, the gravity of stress and exertion was lifted away. The masseuse delved deeper, coaxing the body into releasing its hold on tension and allowing a flow of energy that revitalized and refreshed. As the massage progressed, the boundaries between the physical and the emotional began to blur. The touch transcended physical relief; it became a gesture of care and connection, a tactile dialogue of comfort, support, and intimacy. The couple found a rare moment of absolute presence with each other, a pause in the whirlwind of life where the only demand was simply to be. Payman reached out and took Bella's hand without a word between them.

The session concluded with a gentle decrescendo, the hands of the therapists retreating like the ebbing of a tide. The couple was left alone to bask in the afterglow of the treatment, their bodies relaxed, their spirits intertwined. They slowly rose, their movements languid and unhurried. Payman found their warmed robes. He lovingly wrapped one around Bella then covered himself in the warmth of his own robe. They stepped out into the cool evening, the air crisp against their rejuvenated skin. As they proceeded to a patio, the warmth of the hidden heat lamps and the heated floor making any chill a fleeting memory, an attendant handed them a crisp but fruity champagne and escorted them to comfortable, plush seating adorned with warm blankets and soft, glowing lanterns scattered around for a touch of warmth and intimacy.

As they explored the culinary appetizer treats, Bella noticed a personal note from the chef elegantly placed beside a plate of delicate chocolates. The message was personal to her, welcoming her and inviting them to relax and savor the moment. Bella accepted this welcoming reminder of both her perfect day and to 'be in the moment'. She smiled and passionately kissed Payman. Very much in the moment and very appreciative.

The first course was a beautifully presented amuse-bouche of a veal dried peach carpaccio, a tantalizing hint of the chef's creativity, featuring local ingredients that capture the essence of the Alps. Each course was paired with a perfectly complementary wine. Many of the wines were local with surprising perfection.

The main course was a local version of the surf and turf, a perfectly cooked piece of venison and mountain trout, accompanied by a selection of seasonal vegetables, each showcasing the flavors of the region with a modern twist. As the meal concluded, a stunning dessert was served, combining the sweet and the sublime in a dish that featured rich chocolate, local berries, and a hint of Alpine herbs, accompanied by a dessert wine.

As the perfect evening wound down, the two comfortable on a perfectly placed loveseat, Bella nuzzled into Payman's neck. "Payman, I love you. This day has turned out perfectly."

"Bella, I love you too," Payman ran his fingertips through Bella's hair. "I'm glad you held me to a massage. It's exactly what I needed, it put me in the perfect state."

"I don't want to move, I'm so comfortable in this robe, under these blankets and wrapped in your arms."

"There is no rush, but we'll have to give the chef this fabulous table back at some point. You can stay wrapped in that warm blanket. I've made arrangements for us to get picked up at the doorstep. Our clothes are packaged and stored in the car. We can cuddle up in the backseat."

After the two were ensconced in their ride, Bella thought for a moment, nestled against Payman. "I don't think I'm ever going to let this blanket go." Bella joked as the Range Rover whisked them away.

CHAPTER 29
DISAPPEARANCE

Crans-Montana Swiss Alps, Switzerland
19 March 08:20 Local Time

Bella slowly awoke to the household coming to life—the aroma of freshly brewed coffee and French toast After skiing hard for five days, she had slept for ten hours and checking the clock, decided she deserved a day off.

"Wake up sleepy-head. It's your boyfriend with breakfast," Payman announced. "I know you wanted to sleep in but I wanted to say goodbye before taking the helicopter to ski the back bowls."

"Oh, you're sweet," Bella said, lifting herself up. "I'm so groggy, I just keep falling back into dreams." Payman set a tray on the bed with coffee, French toast, fresh fruit, and a chocolate. He gave her a kiss and said, "I'll be up with the guys you've met from my grandfather's security detail. They live here and know the mountain. Don't worry. Have a relaxing day. Enjoy the sun and a book on the deck. See you for dinner."

Bella hugged him and said, "Be safe. It can be dangerous in those back bowls. Stay away from the snow built up on the cornice. I know you know what you're doing but I have to remind you that's the job of a caring girlfriend."

"What will you do today?" Payman asked. Bella grabbed him by the shirt and kissed him.

"Something lazy," she said, her eyes dropping slightly.

Bella found it an extraordinary day for shopping in Crans-Montana. It was peak ski and tourist season and thus peak shopping season. The pristine snow blanketed the landscape and the sun glinted off the white peaks, casting a magical glow over the bustling resort town. Vibrant streets invited Bella with their elegant charm, a fusion of chic sophistication and traditional Swiss allure, alive with the excitement of the winter season, and lined with high-end boutiques, luxury watchmakers, and exclusive fashion houses.

She wandered down the promenade with a clear, free mind, observing the enticement of each store with its beautifully decorated windows showcasing the latest in haute couture, cutting-edge ski gear, and exquisite jewelry. She found herself drawn into a ski boutique that advertised collections of designer ski wear. Here, a personal shopper offered her service. She didn't need a new ski outfit, but what the heck. She was drawn to a technical blue puffy, the color of Payman's eyes. It was matched with off-piste technical powder bibs. Being an expert skier, Bella chose the new ski outfit more for its functionality than for its fashion.

Next, she discovered a local artisan shop, where handcrafted goods told the story of the region's heritage and craftsmanship. Unique pieces of jewelry caught her eye but she decided none of them were perfect for Payman.

For a casual lunch, she enjoyed an amazing salad on a terrace overlooking the many ski runs, entertained by a handsome waiter that shared her enthusiasm for both skiing and mountaineering. He indicated an equal interest in her, but she made it clear she was taken. It made her proud and satisfied to declare she was taken. She bid the handsome waiter a good day and wished him amazing skiing.

She continued exploring, stopping at a luxury watchmaker's shop filled with exquisite timepieces. A watchmaker saw her delight with the watches. She introduced himself as Denise and asked in French if she could be of assistance. Bella answered slowly, and Denise smiled.

"What kind of person is he?" she asked.

"Who?"

"The man you are buying this for." A devilish look appeared on her face at Bella's confusion. "Nobody is as careful about their choices as you are unless they are making a choice for an unsuspecting paramour."

"Am I being that obvious?" Bella laughed.

"You're still young. You are not so jaded that you hide your passion. Mademoiselle, if you are here for your lover, then tell me: what are you looking for? Describe your boyfriend."

Bella collected her thoughts. "He's a world-class sailor, wears a Yachtmaster given to him by his grandfather…so he appreciates Rolex. He is a risk taker, I know from tree skiing with him yesterday. Today he's in the back bowls heliskiing. He's very smart, a PhD. Physicist. He's ruggedly handsome, Persian exotic with Mediterranean blue eyes, in great shape…" She took a breath, about to continue.

"I see," Denise said with a smile. "We don't have a large inventory of new watches and those are standard styles. I'm happy to show them to you, but I know your idea of a gift is beyond what I can offer as a new Rolex. I think I have just the watch for you, a meteorite dial, platinum Sky Dweller."

Denise gestured to a dispenser, offering Bella complimentary beverages. Bella helped herself to some mineral water. As she sipped and took a cleansing breath, she managed her Zen—she was in the moment.

A few moments later, Denise returned with the timepiece in hand. "This is an extraordinary timepiece, truly a one-of-a-kind. I am in love with this watch myself. The Rolex Sky Dweller displays two time zones and features an annual calendar. In platinum with a meteorite dial, it is the pinnacle of luxury and craftsmanship, merging the art of watchmaking with the ethereal beauty of celestial materials. The Sky Dweller not only signifies time but also carries a piece of the cosmos on your wrist."

Bella was breathless. Denise with gloved hands gently slips it on Bella's wrist. "I'll bet it feels wonderful, and it looks perfect on you, I'm sure your lover will love it too and appreciate its uniqueness." Bella stared, transfixed by the timepiece.

"It's perfect," she said simply. She had the watch wrapped and told Denise of the couple's trip, to which Denise listened in rapt fascination. Bella thanked her and left the shop with a burst of confidence, quietly amazed at the wisdom of the woman she had just spoken to.

She perused the local art galleries before the sun began to cast a golden light over the mountains. With treasures in hand, she meandered away from the market street, returning her senses to the mountain air. Truly Zen.

Bella bypassed a ride back to the chalet—the brisk walk back would satisfy her appetite for daily exercise. The beauty of the mountains and the brisk air was refreshing. The chalet came into view as the alpenglow descended on the mountain. With her heart rising she noticed a fleet of range rovers in the circular driveway as she approached the chalet. She entered the chalet and was immediately met by Payman's grandfather, Amir. Who without a word came over to her and wrapped his arms around her in a loving hug.

"I have been trying to reach you all afternoon," he said, his low, monotonous tone causing her to slow her tell me breath as he gradually released her.

"What has happened?" she asked. "Has something happened to…?" Her throat swelled; the words barely came out in a whisper. Amir placed an arm around her shoulder and instructed her to sit. He handed her a glass of water, the delicacy of his actions giving a mournful, almost ceremonious feel. Bella remained quiet, all neurons firing to navigate the rising dread.

"Tell me, Amir! I can't bear it any longer. What's happened?" Bella blurted out, both hands around the glass, sweat building in her palms at the sight of the tears in Amir's eyes.

"I got a call from Alan, my chief of security. Payman and two members of the security team had gone skiing, but. ." he swallowed deeply. "We lost track of them."

A long gasp escaped Bella's throat, "When?"

"Around two hours ago. Alan contacted emergency services. It was the second the lost signal was confirmed, and search and rescue are already looking for them, but…we don't know where he is."

Bella's hands shook as she took a drink. "Are the helicopters still flying? I could get my gear and help. Maybe follow the same route!" Tears flowed from Bella's eyes, she was in shock. She was shaking from toes to shoulders.

"We must let search and rescue do what they can. Our best approach is to wait here and pray for their safety. The couleurs and crevasses are dangerous after the snowfall."

"You want me to pray?! He might be already dead!"

"My dear—"

"Amir, we must do something. He might be…"

She put the glass down. Their faces were now sodden. Amir reached for her and held her to his chest, promising that it would be okay.

After hours of waiting, the sky had gotten dark; the search was called off for the night. Bella agreed to some light sandwiches but remained secluded in the bedroom. Her gifts were neatly arranged, the watch sitting atop the rest of them. She stared at it, grieving for the excitement she had in anticipation of presenting it to Payman. As physical fatigue set in, she laid down fully clothed and cried, barely blinking with her eyes on the gifts, until she, at some point, fell asleep.

The new day was cold and dreary. The wind picked up and clouds descended on the mountain. There was no good news from the search. It was all but confirmed that Payman vanished on the mountain.

Bella was numb, as was Amir. There was little verbal exchange between them but their much-mirrored demeanours was all they needed. Bella, needing something to do, packed her belongings without even being aware. On her wrist was the Rolex meant for Payman.

They rode in silence to the airport. The flight in the luxury private jet back to their homes in England was just as silent, interrupted only by Bella's occasional sobs.

CHAPTER 30
BEA & BELLA

Bentley Hall Oxfordshire, England
21 March 11:40 Local Time

Upon her return from Crans-Montana, Bella remained under the care of the family physician who made frequent visits to Bentley Hall to ensure his patient was appropriately grieving and had any medical support she may need. Bea took it upon herself to shield Bella from unwelcome visitors, carefully curating her company and vigilantly overseeing her recovery. If it had been entirely up to Bella, she would have buried herself beneath the covers indefinitely, shutting out the world and the sunlight that intruded upon her sorrow.

It was the second time Bella had returned to Bentley Hall with a weight on her heart, the first being the life-threatening injury at sea and long recuperation away from Bella in his native country Iran. But this was different. Bella had fully given herself totally to Payman and he had done the same. Their vacation had felt like an unspoken commitment to their future together, a future that now seemed heartbreakingly out of reach.

On an unusually sunny first-day of spring, Bea coaxed Bella to sit with her on the patio hoping the warmth and renewal of the Oxfordshire countryside would bring some solace. After the chill of winter, the air, no longer biting, carried the soft whispers of emerging life. The morning had started with a soft mist, shrouding the rolling hills and ancient, ivy-clad buildings of the estate, and as the sun rose higher, the fog thinned, revealing a landscape poised on the brink of renewal. Birds chirped in the distance, their songs signaling a return from winter migrations, while the rolling hills glistened in the fresh light.

The chef had prepared their choice: an Asian shrimp salad. Plump, juicy red shrimp tossed with a mix of crisp greens, romaine and arugula providing a fresh, crunchy base. Bright slivers of almonds, vibrant red bell pepper, thin slices of cucumber, crunchy wontons and a scattering of shredded carrots. The colorful dish was as much a feast for the eyes as it was for the palate.

As the tantalizing aroma reached Bella, tears welled in her eyes. The scent pulled her back into a tide of memories: shared meals with Payman, stolen laughs in the kitchen, and quiet moments at the table. She smiled weakly, a bittersweet expression of gratitude and loss. Though she gave a melancholy smile, the thought of never sharing such a meal with Payman again was another moment of loss. She took a deep cleansing breath, knowing she needed to move forward. The food did smell and look delicious. She took careful bites and did her best to savor the taste, her mind and heart too broken to really think about what she was doing. The flavors were exquisite, but her heart was too heavy to savor them.

"Thank you," Bea told her. "Just for sitting with me, sweetheart." Bea's voice was soft, understanding Bella's reluctance to respond. She let the silence linger before continuing. "We all feel your pain. Your uncle, the staff, and I—we were beside ourselves with worry when we heard the news."

Bella lowered her gaze to the plate. Her voice was barely audible. "We were building something together. We had it all figured out..." Her words trailed off as she caught herself, realizing the futility of their plans now. A bitter thought crossed her mind: They had figured it out like researchers drafting a hypothesis—methodical, careful, and fragile. She stopped herself from speaking further as a thought came to her, one she didn't have the humor to entertain: they had worked it out like the researchers they were.

The dappled sunlight played across the table, casting shifting shadows through the newly budding trees. The soft rustle of a breeze mingled with the faint clink of ice in their glasses, filling the quiet space with a sense of stillness.

Bella finally spoke again, twirling her fork through her salad with a distant look. "There's something special about the first sunny day of spring," she murmured. "I'm just not sure what I should do with it."

"You don't have to do anything, darling," Bea replied with a small, reassuring smile. "Sometimes, after an unexpected loss, the best thing you can do is simply allow yourself to feel."

Bella remained still, let the words settle, a wave of grief fall through her. After a while, she allowed herself to eat a little more. She took another bite and looked at Bea. "Would you like to go for a walk?"

Bea smiled softly, nodding. "I'd like that."

They took themselves across a lush, verdant winter grass lawn that stretched out like a green carpet underfoot. The countryside was alive with the subtle sounds of spring: the distant calls of returning birds, the gentle rustling of young leaves in the breeze, and the occasional buzz of an early bee exploring the first blooms. At the edge of the lawn, a shaded path took them through a copse of winter trees, their branches forming a loose canopy overhead. The ground underfoot became a mix of soft earth and fallen leaves from the previous year. The air here was tinged with the scent of moss and wood—a deep, earthy aroma filled the lungs with each breath. Bella and Bea quietly lost track of time and, through the thickness of the trees, they came to a winding path that ended at a quiet lane. People strolled and cycled by, moms and dads with prams enjoyed the milder air. Couples walked hand in hand, sometimes snuggling. Bella noticed but was determined to move forward.

"We're not far from the Globe," Bella said. "Let's stop in and see Arthur."

Bea replied, "You know you don't have to twist my arm to raise a pint." Bella gave a stuttered breath, something close to a chuckle.

"Your uncle introduced me to the Globe on an early visit. I think it was there that I first felt a warmth developing in me as I watched him so comfortably mingle with the locals. There were no airs about him. Let's go."

As the Globe came into sight the two caught one another's eyes and chuckled in anticipation of a favorite experience.

They neared the many centuries old front door, barely one hundred fifty centimeters tall. All the locals knew to "mind your head" crossing the threshold. The smell of aged wood and the faint tang of spilled ale greeted her, mingling with the low hum of conversation and laughter that felt like a balm to her frayed nerves.

Then she heard it, a friendly cheer from the regulars, "Oy, young Bella, daughter of beautiful Mary, mind your head!"

Bella again heard the usual greeting from the regulars. Hearing their familiar voices brought a bittersweet ache to Bella's chest. It was as though their warmth reached out to fill the void Payman had left, even if only temporarily. She smiled as she caught the gaze of so many of her friends from the village and greeted each with a hug and often a cheek kiss. For the first time in weeks, Bella felt herself relax, allowing the warmth of familiarity and kindness to momentarily lighten the weight she carried. Then, taken by the rush of serotonin, or the happy bodily chemical, she turned to the bar.

"Arthur, come out from behind that bar and let me give you a hug and a kiss on that scratchy beard."

Arthur didn't say a word until they were both embracing. "My girl," he said. "You know I'm so sorry for the loss of your fella."

Bella thanked him sincerely. When Arthur asked her how she was doing, she gave no response, simply tightening her hold around him. When they finally let each other go, Arthur marched back around and pulled a pint. When Bella reached into her purse, Arthur raised his hand to her.

"Sorry, young one, this round is on the house for all, just to celebrate your return. It's always nice to see family." Arthur chuckled.

Bella, defeated by the kindness, accepted the drink and allowed herself to bask, if only for a little while, in the graciousness and decency of the Globe. She mingled, chatted up the regulars and laughed. Bea was so relieved to see her more comfortable again. It must be the magic of the people of the Globe. Bella texted Uncle Andrew and invited him to join them.

By the time he arrived, the sun had already sunk far below the horizon.

"It's good see you up," he told her. "Don't think I've ever seen you beaten in your life."

"I might have been more ready for this than I had realized," Bella said. "I'd be lying if I said I hadn't missed place."

"The panacea for human duress." Andrew gave a wry smile. "You know what else can't hurt?"

He turned, as if he'd orchestrated the moment, as Arthur came out to assist his daughter, Abby, brought out the dinners—traditional fish and chips- golden, flaky white fish in a light beer batter with minted peas and thick-cut chips drizzled with vinegar. Arthur carried fresh pints. All was well.

They ate in comfortable silence, Bella rediscovering a little of the pleasures of the food. Despite large periods of quietness, she could talk a little more, and the trio mingled, hugged, laughed, and clicked glasses with the Globe's patrons.

As evening descended upon them, they had bid a good night to the Globe and Bella had hugged Arthur thanking him for, in her words, just being him. Andrew had arranged with Charles for them to be picked up. They settled into the plush leather of the seats. Bella closed her eyes and sighed. Her mind cleared with brief content as the soft purr of the BMW's engine soothed her. The lights of the town diminished as they transitioned through the forest, and Bella's want of homely comforts renewed when the castle appeared beneath warming spotlights.

After she had gotten out of the car, Bella hugged Bea. "Thank you."

"You okay, sweetheart?"

"No…but that's okay. I'm much better after the day we've had together."

"We're still here," Bea said, stroking Bella's hair. "You take all the time you need. Bien?" Good?

Bella sighed, smiled, and replied, "Si. Todo bien. Contado." Yes, all is well. Perfect.

CHAPTER 31
RETURN TO WORK

Oxford University, Oxford, England
End of March

After her period of rest and recovery at her family estate, Bella immersed herself in the challenges and complexities of her work, examining the atom from the perspective of the smallest particles of the quantum atom. Bella's teammates were very supportive and welcomed her with open arms and hugs. She fell into her routine of Saturday afternoons at Blackwell's Books and some evenings at the Turf Tavern. Her research was progressing smartly and expediently with some recent discoveries that placed her and her fellow researchers in a position to be invited to conduct their experiments at CERN.

One evening at the Turf, Bella was sharing pints with her colleagues. They were celebrating the invitation to CERN.

Jenni, one of Bella best friends and a fellow researcher, raised her glass and said, "Bella, we're so happy you're back. And together, we've been accepted to CERN. It's truly a geek's dream vacation." Pints were raised, clinked and ale downed.

Bella laughed, "Yes a geek's dream, CERN. We are going to solve the questions of the details of the universe. I look forward to examining the duality principle Einstein proposed. I have a new concept I'm studying. I could only do such research from CERN. But for now I'm ready for meat pies." The group celebrated as their food order arrived. Bella had sampled the Turf's selection of pies: beef, variations of fish, lamb, pork, and chicken. These fish or meat pieces were slow-cooked with a mixture of onions, carrots, and mushrooms, bathed in a thick, robust gravy seasoned with herbs like thyme and rosemary, and occasionally a splash of stout or ale to deepen the flavor.

The group followed Jenni's lead and ordered a fresh round of pints to accompany the pies. They then enthusiastically engaged in a discussion about CERN, as only a collection of geeks could do. CERN, the European Organization for Nuclear Research, located near Geneva on the Franco-Swiss border, was founded in 1954 with the aim of providing a facility where European scientists could perform cutting-edge research in physics in a collaborative, international environment. Among its various facilities, the Large Hadron Collider (LHC) was the most famous. The world's largest and most powerful particle accelerator circled a 27-kilometer underground tunnel. It was designed to collide particles at close to the speed of light, allowing scientists to observe the resulting interactions between particles that are created in these high-energy collisions.

Beyond the LHC, CERN operated a host of other smaller accelerators and experiments designed to explore different aspects of particle physics, such as the study of antimatter and the properties of neutrinos. CERN is also a leading center for developing and testing new technologies for accelerators and detectors, and many technologies developed here have found applications in other fields of science, as well as practical applications in medicine and industry.

Notable not only for its scientific endeavors but also for its commitment to operating as an open-access laboratory. CERN is supported by 23 member states, and scientists of all nationalities collaborate on its experiments. Moreover, CERN has been a pioneer in the digital arena; it was here that the World Wide Web was born in 1989, invented by Sir Tim Berners-Lee as a tool to facilitate information sharing among scientists participating in experiments at CERN and around the world.

CERN served as a beacon of scientific inquiry and collaboration, where fundamental questions about the universe were explored and where significant contributions to technology and society were regularly made. Bella was especially excited with anticipation of her opportunity to find support of Einstein's duality. The group discussed their upcoming trip with the glee of kids going off to a favorite summer camp.

Jenni turned to Bella and whispered, "I know you are hurting and you miss him. I'm here to listen and as a shoulder to cry on if you need me." She smiled at the sight of Bella mildly blushing. "You're our Niels Bohr, smart, good with people, and beautiful. We're really glad you're here with us."

Bella hugged Jenni. "You have been here since our first meeting at the Physics Club when I was a fresher and joined you all at the Bodleian. I was so out of my league that day."

"You? You've never been out of your league. Some of us were intimidated, but maybe that was our arrogance getting the better of us," Jenni laughed. "You know we were always cautious of Payman. He was so mysterious, never really shared his past. It was you who brought him out, and I learned to appreciate him."

Bella took a long, long drink, composing herself. "It's been horrible. Everything I've done and experienced has seemed to be done through some refracted lens where everything points to him. This can't continue. Now the challenges of our work come home to roost. I must focus. I'm turning the corner, I'm putting myself into our work and together we will expand the knowledge of the universe." Bella raised her glass again. "To the universe." Jenni cheerfully obliged her. All clinked their glasses with a smile.

A young man, undoubtedly a student, approached Jenni, the tutor of Quantum Physics, "Professor Hobbs," he said. "My name is Bryce. I'm taking your Quantum and Relativity class."

Taken slightly aback, Jenni put up a front to stop herself from giggling but admiring his boldness. "…hello Bryce."

"I have a couple of questions. Would I be intruding to ask them here?"

Jenni brought his attention to an empty seat with an overly polite gesture. "You've come to the right place. We have eight of the world's leaders in quantum here at the table, almost a Solvay conference. Don't be intimidated. None of us would be tutors if we didn't love questions."

Bella beamed her best connecting smile, then piped in, "I promise you, we're actually a very friendly bunch."

Bryce, still unsure but slightly more relaxed, asked, "In class, you said, 'God does not play dice with the universe.' I've been fascinated by that comment. My mates and I have been discussing it. But I'm struggling to understand it."

Jenni smiled. "As much as I'd love to take credit for that quote, it's actually from Albert Einstein. He said it to express his doubts about the randomness in quantum mechanics." She paused and added with a grin, "Mark, care to explain while I enjoy my drink?"

Eager to join the conversation, Mark leaned forward. "Einstein, as you know, is famous for his theory of relativity and his belief in a predictable universe. He thought everything in nature followed a clear cause-and-effect chain—nothing random, nothing left to chance. So when quantum mechanics suggested that some events could only be explained through probabilities, he rejected the idea. To him, this randomness made the theory incomplete. He believed there had to be hidden rules or variables we just hadn't uncovered yet."

He gestured toward Bryce to emphasize his point. "Take light, for example. We see it visually so we logically view light as a wave. But, in quantum mechanics, it behaves like both a wave and a particle at the same time—that is until someone observes it, and then it 'chooses' one form. This concept, known as wave-particle duality, clashed with Einstein's view of an orderly, deterministic universe. Meanwhile, Niels Bohr, another legendary physicist, fully embraced the probabilistic nature of quantum mechanics. Bohr argued that uncertainty and chance are fundamental to how the universe operates."

Pausing to let it sink in, Mark continued, "The debates between Einstein and Bohr were some of the most fascinating in the history of science. They laid the foundation for much of what we understand about quantum physics today. That's why we're so excited about CERN—it gives us the tools to explore these questions further and, maybe, bring some clarity to them."

Jenni chuckled and chimed in. "To sum it up, when Einstein said, 'God does not play dice,' he meant the universe shouldn't rely on chance. Bohr's reply? 'Stop telling God what to do!' And decades later, Stephen Hawking added his twist: 'Not only does God play dice, but he sometimes throws them where they can't be seen.' That's quantum physics for you, Bryce."

Bryce blinked, looked more confused than before. He shyly responded, "I may have to make a couple of clarifications with my mates. Thank you for letting me interrupt."

With a laugh, Jenni said, "Good luck. And don't worry, it's confusing for all of us—that's why we keep researching." Jenni drained her pint.

"And anytime you want to confuse Hobbs, be my guest" replied Mark.

CHAPTER 32
QUANTUM COMPUTERS

Oxford University Oxford, England
Early April

Bella and her intrepid group of physics club members had settled into dream jobs, researching, teaching, and writing about their favorite subjects in physics. Quantum remained a favorite topic among them. One night, they gathered for dialogue and debate—not so much in a formal club setting but rather as renowned experts, reminiscent of the way artists and writers met to discuss their craft in the bohemian bars of Paris in the 1920s.

That evening, the group met at The King's Arms pub. Established in 1607, it is one of Oxford's oldest pubs and steeped in centuries of rich academic and local history. Nestled near the Bodleian Library, it has long served as a favored gathering spot for scholars, students, and tourists alike. Its warm wooden interiors and low-beamed ceilings preserve the feel of a traditional English alehouse. The pub is renowned not only for its historic charm but also for its role in Oxford folklore and its enduring connection to university life—frequented over the years by notable figures such as philosopher Isaiah Berlin, physicist Richard Dawkins, and writer J.R.R. Tolkien.

Jenni began with a comment she had been reflecting on, inspired by the 14th-century poet and philosopher Dante. "To retranslate Dante's famous line from the Commedia, 'Abandon hope all ye who enter here,' I'd say, 'Be enlightened by hope all who enter here.' We've learned that the only way to approach our science is to be prepared to be enlightened when we enter the world of quantum."

Mark smiled and chuckled. "Brilliant, Jenni. It sounds like you'll be tutoring philosophy in the Oxford summer school next July."

As pints and pizzas were consumed, the lighthearted banter shifted to the topic of quantum computing.

Ernie, one of the newer members of the cohort, wiped away a dribble of beer that had missed his mouth and was hanging from his beard. "The concept of quantum is as challenging to us today as the idea of flight would have been to a 14th-century philosopher. Imagine telling them that humans would one day leave the planet, fly through space, and return to tell the tale. Just as space travel would have been incomprehensible to their minds, quantum mechanics remains equally unfathomable to many in the 21st century. Yet unlike them, we don't have to wait 500 years to see it unfold."

He continued, his tone growing more passionate. "Quantum doesn't operate in a black-and-white, digital world. It exists in a world of brilliance, where all colors are experienced at once. The binary on-or-off structure of digital computing doesn't apply.

Quantum is both, neither, and everything at once—it's the brilliance of the universe, available to us today."

Quantum physics, he explained, explored matter at its most fundamental level—a world within our world, built from the basic building blocks of nature. It was unimaginable to laypeople but lived in the minds of theoretical explorers.

"As our quantum theories begin to take shape in reality," Ernie added, "we're going to need a completely new approach to computing."

The group agreed. Quantum mechanics offered a rich computational model that delved into the physical world, allowing scientists to visualize and build quantum computers and networks. They reminded one another that these were systems based on quantum mechanics at the most fundamental level.

"So, as quantum physics and methodologies develop at the atomic scale," Ernie said, "our ability to rethink data pathways and how that data is processed will evolve as well."

At its simplest, he explained, today's computers relied on the movement of electrons in silicon chips, using a structure of zeros and ones—digital, either/or. But quantum computing required entirely new materials, functioning at the smallest components of the atom. These materials were being explored by scientists at the world's leading institutions.

"One of the oddest tenets of quantum theory," he continued, "is that a particle can be in two places at the same time. Applying this to quantum computing, its exponential power comes from qubits, which can exist as 0 and 1 simultaneously. This allows quantum computers to process information 158 million times faster than today's supercomputers. They can accomplish in minutes what would otherwise take millennia. It's all thanks to two key principles: interference and entanglement. Interference cancels out unwanted solutions while enhancing correct ones, and entanglement ensures that the combined state of qubits contains more information than the individual qubits alone. Together, these principles have no classical analogy. For instance, 2 qubits are equivalent to 512 bits, while 16 qubits equal one megabyte."

Ernie referenced a 2015 Stanford University experiment. "Physicists demonstrated that a group of atoms in superposition could exist in two places at the same time. This means that when an atom is observed in one place, it can also be considered partly in other places. The original state must be regarded as a superposition of multiple states. While challenging to grasp, this concept could lead to breakthroughs in quantum computing, particularly in harnessing superposition."

Quantum computing, he concluded, was a rapidly emerging technology harnessing the laws of quantum mechanics to solve problems far too complex for classical computers. Even today's supercomputers struggled with tasks like modeling the behavior of individual atoms in a molecule, where countless electrons interacted. Quantum algorithms, however, approached these problems by creating multidimensional spaces where patterns emerged.

"Quantum computers are elegant machines," he said. "They're smaller and require less energy than supercomputers. They will transform how we work in research, especially in areas like modeling chemical reactions or protein folding. This is the future we'll witness in our lifetimes."

After this heavy theoretical discussion and debate, the intrepid group of intellectuals let out a collective sigh of relief. The challenges were daunting, but they were ready to tackle them together.

As the night wore on, the group's mood lightened. They ordered another round of pints, laughing over shared memories from their university days and trading anecdotes about experiments gone hilariously wrong. The dense fog of theoretical debates lifted, replaced by a shared sense of purpose and excitement for the possibilities that lay ahead.

Jenni raised her glass. "To quantum dreams and daring experiments," she said with a grin. "Here's to solving the unsolvable."

"To the unsolvable!" they echoed, clinking glasses with a cheer.

Bella leaned back in her chair, gazing at her friends, her thoughts straying for a moment. She saw in them the same spark of curiosity and determination that had ignited their journey years ago. Despite the daunting challenges ahead, there was comfort in knowing they would face them together.

"We may not have all the answers," Bella said softly, "but isn't that the beauty of it? To explore the unknown, to make the impossible possible?"

Ernie nodded. "And to leave something behind for the next generation of dreamers to build on. After all we recognize we don't know everything there is to know."

The group fell quiet for a moment, the weight of their shared mission settling in. Outside, the sounds of Oxford nightlife carried on, but inside the pub, their world was one of ideas and discovery.

As they said their goodbyes and stepped out into the cool night air, Bella looked up at the starry sky, the same one that had inspired countless scientists and philosophers before them. Somewhere out there, she thought, the answers they sought were waiting. All they had to do was keep asking the right questions.

With a renewed sense of purpose, they walked together into the night, ready to face whatever challenges the quantum universe had in store.

CHAPTER 33
CLASS

Oxford University, Oxford, England
08 April 14:00 Local Time

Bella and her colleagues dined, drank, and debated physics at their favorite location, the Turf Tavern. As the warm glow of conversation filled the room, Jenni noticed Bryce approaching the group with a slightly more confident stride than before. She chuckled, reminded of the time Bryce had attributed Einstein's famous quote, "God doesn't play dice with the universe," to her by mistake.

With an endearing smile, Bryce reached the table and turned his attention to Jenni. "I'll bet it's not often a student says this, but I can't wait for class tomorrow. I hear we might have visitors who could help explain Stephen Hawking's comment about God throwing dice where we can't even see them!"

Bella smiled warmly. "Bryce, we may be one of the most exciting collections of physicists since Solvay, but I doubt we can compare to Einstein, Bohr, or Hawking. Still, I'd agree—Dr. Hobbs' class will certainly be worth attending. I hope all your questions get answered." She added a discreet wink.

The next day, Jenni held a lecture in one of the traditional lecture halls. Dark wooden paneling, intricate moldings, and leaded glass windows that had borne the crests of various colleges. The seating arrangement, wooden benches arranged in tiers faced forward toward the main focal point of the hall—the lecturer's podium. Projectors, speakers, supported contemporary educational practices, subtly integrated to minimize disruption to the historical aesthetics, save for the large projection screens behind the lecturer's podium.

Jenni reviewed her lesson plan for the discussion of quantum as students filed into their seats. She braced for the glazed-over eyes of the undergraduates, determined to make sure there was discussion and maybe an understanding that would challenge the students to further explore quantum. To quote the great physicist Niels Bohr's, "If you think you understand quantum mechanics, you don't understand quantum mechanics." Knowing that the best she could do was to engage the students and spark some interest to question, she stepped to the lectern.

As Jenni raised her view to assess the audience, she was pleasantly surprised to have seen Bella and Mark, her two fellow colleagues, front and center, awaiting the coming discussion. She kept a clicker in her pocket, which she kept her hand firmly around. She clicked it once, and the screen behind her brightened with the words:

****Introduction to Quantum Mechanics: Unveiling the Quantum World****

****Learning objectives for today:**

****Understand the fundamental principles of quantum mechanics****

****Introduction to Key Contributors****

****Recognize how quantum mechanics differs from classical physics****

"Today, we embark on a journey into one of the most intriguing and profound areas of modern physics—quantum mechanics."

As her eyes flickered around the room, she noticed the familiar mix of expressions on her students' faces: strained focus, calmness, and either dread or exhaustion. It seemed one of them might've been hungover.

"This field not only challenges our classical intuitions about how the world works but also underpins much of the technology and theoretical frameworks in contemporary science. Do not be surprised or slowed by not understanding something. We will spend the time necessary for us to progress into the world of quirky atoms." She smiled and awaited a reaction. Seeing none, she moved on.

"I must acknowledge two of my esteemed colleges, sitting front and center. Dr. Bella Arensico, a world leader in research on entanglement in quantum systems and Dr. Mark Douglas, equally as qualified as a contemporary quantum expert. We may call upon him to help explain duality. I hope they are here to learn and not heckle."

"No promises," Mark said. A few murmurs of polite laughter spread through the room. Jenni continued, clicking away to produce images on the screen.

"I'll begin with an example of a concept we all know well: the X-Ray. The X-Ray was first observed by German scientist

Röntgen who found them by accident as he was experimenting with early vacuum tubes. He took a photograph of his wife's hand revealing her bones and the wedding ring surrounding her finger. Röntgen nor anyone else at the time knew what caused this. He referred to this unknown variable as "X" for unknown radiation, just like we may use X in a formula for an unknown variable. Years later, we were able to describe this process in detail and make the X-Ray a common procedure."

At that, a hand shot up near the back. Jenni, pleased at the early engagement, nodded to the student. "Yes, Art?"

"May we use the "X" as an answer on our exam if we don't know the answer?" Several chuckles popped across the room.

"Yes, Art, you may use the "X" for an unknown variable or unknown concept. However, to get full credit be sure you describe what you know and do not know about the concept and how it relates to the learnings of this class."

Jenni's cheeks puffed as she continued to chuckle, and she proceeded with the next slide:

****Historical Prelude****

"Quantum mechanics didn't emerge in a vacuum. At the turn of the 20th century, classical physics faced several puzzles they could not solve: the stability of atomic structure, the nature of light, and the behavior of materials at microscopic scales baffled the brightest minds. Max Planck's introduction of the quantum hypothesis in 1900 marked a radical departure from modern physics. He proposed that energy is quantized, meaning it can be exchanged only in discrete amounts, which he called "quanta".

For those who need a refresher, a blackbody is a theoretical object that perfectly absorbs radiation at all wavelengths and thus, when in thermal equilibrium, also emits at all wavelengths. An almost closed cavity with only a small hole through which radiation can escape is a good practical approximation to this ideal. It absorbs all incoming radiation and in turn emits the same amount of energy. Remember back to Newton and the conservation of energy?

A simple visible example of a blackbody would be the iron rod a blacksmith puts into a hearth of coals to work the iron into a work of his art. The blackbody is the iron rod. The blacksmith inserts the rod into the furnace, the rod absorbs radiation in the form of heat energy radiated in the furnace. As the rod absorbs more and more of the energy, it increases in temperature, the observable energy absorbed changing the color of the rod from yellow to red, to blue, to white. So, this black body rod absorbs the radiation in the form of heat from the furnace and gives off radiation in the form of visible light radiation and heat.

In the early 20th century, this blackbody concept resulted in unexpected results when the intensity of ultraviolet radiation decreased with wavelength rather than increasing as predicted by classical theory. It was dubbed the 'ultraviolet catastrophe', one that stumped classical physicists. This catastrophe highlighted a fundamental flaw that called for a new way of looking at and understanding the microscopic world.

Then comes the ingenuity of Max Planck to the rescue. Our friend Max resolved the catastrophe by proposing that electromagnetic energy could be emitted and absorbed only as a quantized form, discrete packets, introducing the concept of quanta. These are particles of light or photons. It was necessary for the early 20th-century physicists to understand how something expected to be a wave could be a particle."

Jenni was sensitive to her audience, had expected to see glazed-over eyes, but to her surprise, she saw them attentive. Their collective eyes focused either on her or the screen behind her. Sharing a blush with Bella and Mark, she clicked on - the screen behind her displayed the next slide:

****Core Principles of Quantum Mechanics:**

Wave-Particle Duality**

"Bear with me and accept for now that everything in the quantum world can exhibit both particle-like and wave-like properties. Light, which we had understood as a wave, was shown to also behave like a particle in the photoelectric effect—a discovery that won Einstein the Nobel Prize in 1921. Conversely, materials traditionally understood as particles, like electrons, exhibit wave-like behaviors in experiments like the double-slit experiment. Ah…" A few faces fell concerned, and Jenni flicked her finger in Mark's direction. "Dr. Douglas, I don't suppose you could enlighten the room with the description of a double-slit experiment?"

Mark smiled. "The double-slit experiment is a classic demonstration of wave-particle duality, originally performed with light by Thomas Young in the early 19th century to demonstrate that light behaves as a wave. Later, this experiment was conducted with electrons and other particles, revealing that matter also exhibits wave-like properties under certain conditions."

Jenni nodded, even now impressed as the level of knowledge in her colleagues. "Thank you kindly," she said, then pressed the clicker. "Moving on." The screen behind her now displayed:

****Quantum Superposition****

"Quantum entities can exist in multiple states simultaneously—a concept best illustrated by Schrödinger's cat, which is both alive and dead until observed. This superposition collapses to a single outcome when we measure it, a cornerstone concept that challenges our everyday experiences."

Bryce added a comment that demonstrated he had moved beyond his interest in God and dice, and over the course of the lecture, Jenni felt her nerves relaxed as the students engaged all three experts in the room, often causing a discussion between the three. She glowed in content.

"Before we move on, I digress, prompted by a comment from Bryce that pleased my colleagues and me. It was the quote from Einstein, "God doesn't play dice with the universe." Einstein made this comment to Max Bohr, another of the leading physicists of the quantum breakthroughs of the early 20th century. Can anyone tell me what Dr. Bohr's response was to Dr. Einstein?"

A hand shot up and a young woman named Staci answered, "Bohr replied, 'Who are you to tell God what to do?'"

"Extra credit to Staci, perfect, thank you. Yes, the two were actively challenging one another during lively debates from 1925 continuing to the 1927 Solvay Conference. Here is a photo of the participants of that conference. The greatest gathering of physicists ever."

The early color photograph of the conference attended was displayed on the screen.

Jenni continued, "This iconic photograph captures the attendees of the Fifth Solvay Conference on Physics, held in October 1927 in Brussels, Belgium. Widely regarded as the most remarkable gathering of scientific minds in history, this image brings together 29 of the world's leading physicists, including 17 Nobel Prize winners, at the very forefront of quantum theory and relativity.

Seated front and center is Albert Einstein, distinguished by his iconic mustache and contemplative gaze. To his right sits Marie Curie, the only person to win Nobel Prizes in both Physics and Chemistry. The lone woman represented; Madame Currie had two! Also present are Niels Bohr, Werner Heisenberg, Erwin Schrödinger, Max Planck, Wolfgang Pauli, and Paul Dirac—figures whose foundational work shaped modern physics.

The 1927 Solvay Conference is best remembered for the intense debates between Einstein and Bohr over the philosophical implications of quantum mechanics, particularly determinism vs. probability. The conference's central theme was "Electrons and Photons," and it marked a decisive moment in the acceptance of quantum theory.

The historic setting and the monumental contributions of its participants make this photo a symbolic representation of a turning point in scientific understanding—a meeting where classical physics gave way to the new and uncertain world of quantum mechanics.

More on these participants and their contributions in a future discussion. But understand this was the most influential group of physicists ever assembled or probably to ever be assembled."

The slide was advanced and now displayed:

****Uncertainty Principle****

Werner Heisenberg showed that certain pairs of properties, like position and momentum, cannot both be precisely known at the same time. This isn't due to experimental errors but a fundamental property of nature. This principle tells us that the act of observing fundamentally alters the observed."

As Jenni showed a video demonstrating the concept, she tiptoed over to Bella.

Whispered in her ear "How are you doing?"

"Ready," Bella replied. After the video, Jenni brought up the next slide:

****Quantum Entanglement****

With the class fully engaged, Jenni raised her voice slightly in excitement. "Please welcome my friend and colleague Dr. Arensico."

A few embarrassed claps came from a few of the students as Bella got to her feet. She took to the lectern and breathed in the view of everybody watching her.

"Perhaps the spookiest aspect of quantum mechanics is the concept of entanglement," she began. "Where particles become interconnected in such a way that the state of one, no matter the distance, instantaneously affects the state of another. This phenomenon, Einstein called 'spooky action at a distance'. A simple way to understand is that if Dr. Douglas, Mark and I were entangled—don't make anything of that reference, I'm using him because he is a colleague." Jenni and Mark chuckled. A few in the audience giggled unexpectedly. "Imagine that Mark was at his home in Newcastle, and I was here. Theoretically, if Mark accidentally touched a flame and experienced a burn, I would feel the same sensation.

Don't jump to ESP or any psychosomatic explanations, but you may use the term "quantum entanglement" to describe this reaction. The two of us are entangled. These entanglements have been proven to operate faster than the speed of light! Just to clarify, the point is that determining the quantum state of one of a pair of particles determines the state of the other, no matter the distance between them. Remember we are discussing the quantum level. There is really no corresponding phenomenon at a macro scale. One of the most promising applications of quantum entanglement is in the field of secure communication. Quantum key distribution systems use entangled particles to detect any attempt at eavesdropping."

When asked about her work with entanglement and her experience at CERN, Bella continued, "The early experiments were confined to laboratories where entangled particles were separated by mere meters. Today, we're using satellites in space to demonstrate particle entanglement. I'll be at CERN next month conducting a proof of concept in quantum networking which someday will connect quantum computers in different locations, decreasing power consumption and increasing networking encryption."

With a small pause, Bella gestured to Jenni who took her place at the lectern. "Thank you Dr. Arensico," she said. The next slide appeared on the screen behind Jenni.

****Why Does This Matter?****

Jenni proceeded. "Quantum mechanics isn't just a theoretical curiosity; it's the backbone of modern technology. Without understanding quantum mechanics, we wouldn't have semiconductors, lasers, MRI machines, or even understand why the sun shines. It's also the frontier of emerging technologies like quantum computing and networking which promises to revolutionize how we process information."

She took a breath, and continued, "As we delve deeper into this course, remember that quantum mechanics is a framework that allows us to explore and manipulate the very building blocks of nature at the atomic level. It's not just about understanding the universe; it's about harnessing its laws to drive innovation. I encourage you all to embrace the weirdness and wonder of the quantum world, as it is key to unlocking the mysteries of the universe and driving the next generation of technological revolutions."

Jenni brought the lecture to an end. To her surprise the students erupted in applause. The hall buzzed with discussion. It was a success.

As she, Bella, and Mark huddled at the lectern, a few of the students thanked them.

"How was that?" she asked Bella.

"I was less nervous at CERN," Bella joked. The class member's expressions are harder to predict than particles and waves.

CHAPTER 34
QUANTUM CONFERENCE

International Physics Association Annual Conference
Geneva, Switzerland, 04 August

In the bustling city of Geneva, an air of anticipation enveloped the International Physics Association Conference. Scientists from around the globe gathered to share groundbreaking discoveries. Quantum physics and quantum computing taking center stage, the conference hall buzzed with excitement. Researchers from around the globe prepared to unravel the mysteries of the quantum realm. Bella could almost feel scorched from the intense glow of the ideas and atomic calculations flowing between the participants. Such a collection of great minds in physics she could not find anywhere else in the world. Many other researchers working on atomic breakthroughs, especially for weapons purposes, were conspicuously absent. The scientists of the 'axis of evil'; North Korea, Russia, and Iran were as unseen as the dark matter of the universe.

Bella was taking the stage to present her research on 'Entanglement in Quantum Systems'. Her captivating presentation delved into the intricacies of particle interactions, drawing the audience into the fascinating world where particles could be connected regardless of distance. As equations danced on the screen, the audience marveled at the implications of Dr. Arensico's work for the future of quantum communication.

During the next session, on the other side of the conference hall, Dr. Hiroshi Tanaka passionately discussed his team's breakthrough in quantum computing. The room fell silent as he unveiled the latest advancements in quantum algorithms, showcasing the potential to revolutionize computation. The prospect of solving complex problems at an unprecedented speed left the audience in awe, contemplating the transformative impact on various industries.

During the coffee break, scientists mingled and exchanged ideas that transcended borders. Bella listened in quiet rapture as Dr. Maria Santos, a quantum physicist from Brazil, engaged in a conversation with Dr. Li Wei, a quantum computing expert from China. Their animated discussion bridged the gap between theoretical concepts and practical applications, fostering international collaboration in the pursuit of quantum advancements.

In a captivating keynote address, Dr. Alexander Schmidt, a respected physicist from Germany, took the stage to explore the integration of quantum mechanics principles into quantum computing. His interdisciplinary approach sparked discussions on how the synergy between the two fields could lead to unprecedented technological breakthroughs. The prospect of harnessing quantum phenomena to enhance computational capabilities intrigued scientists eager to unlock the full potential of this symbiotic relationship. His impassioned speech echoed through the hall.

Quantum

The first day of the three-day conference culminated in a cocktail reception and conversations with many of the luminaries in quantum research. Dr. Arensico, Dr. Tanaka, Dr. Santos, Dr. Wei, Dr. Schmidt, and others dominated the room with a discussion on the convergence of quantum physics and quantum computing.

The modern conference center in Geneva boasts panoramic views of Lake Geneva and the Alps. With sleek, contemporary architecture, Bella became lost in the beauty of her surroundings. The alpine lake and the snowcapped mountains brought her thoughts first to Crans-Montana and then to the excitement of her skiing with her father. She was in a wonderful place with the crisp sounds of the snow under her skis, the whoops of excitement, the air rushing by her and the sun on her face—she relived a moment from her youth. It was the best she had felt since losing Payman in a not-too-distant time in these same Alps. She smiled inwardly and basked in the Zen of her moment.

Her reverie ended abruptly when an American-accented voice broke through her thoughts. It had the timber and cadence of a memory. A man raised a proposal of a new crystalline structure that he felt could break the quantum computing barrier and usher in the new revolution in quantum computing. Bella thought to herself, Oh, those Americans. Always boastful, always having the answers and wanting the glory.

Remembering the voice, she slowly turned, not wanting to draw attention. She scanned the group. Like a bolt of lightning, she locked eyes with a pair of hazel-speckled emerald eyes. Through some seemingly unknown impulse, her knees went weak, her breathing increased, her heartbeat faster, and her senses were flooded with the memory of something long ago.

What happened next was a result of a chemical reaction, pure chemistry. The man reached through the group and took her arm. Oblivious to everyone else in the room, he blurted out, "Are you okay, Dr. Arensico?"

Bella couldn't react. She was dizzy and speechless. She allowed him to lead her to a table with two vacant chairs. Others quickly gathered around, flowing like electrons attracted to a more stable shell.

She replied, "Thank you. Long day. I just need to sit down."

Alchemized by the moment, she reached out to hold Jack's forearm, which was reaching across the table towards her. "I'm okay. Really." She allowed her bearing to return to her, paying attention to the touch of the man on her arm. "Do I know you?"

The man stuttered, his eyes not leaving hers as if the rest of the room had melted away. "I'm Jack."

"Huh," Bella said, her mind slowly coming back to her. "I think I know that." She took Jack's other hand as he helped her to her feet. For a second Jack was taken aback, but a softness in his eyes told Bella that he was beginning to understand what she was driving at. Then he smiled. "Did we just experience a 'wormhole' moment? Did we instantaneously move between the Camino and Zurich?"

Bella burst into laughter, the others around them exchanging awkward glances as the two of them chuckled, raised from their seats and moved to a high top standing table near the bar. Bella explained to a few others that she would return to them as soon as she could, that she was catching up with an old friend. They talked about their lives, their work, and how they ended up here. No one around existed, no one bothered them, no one interfered.

Jack asked Bella if she had plans for dinner. She told him, "Well, first I need sparkling water. Then, no, I have nothing that couldn't wait. What do you have in mind? A discussion of wormholes?"

Jack's hopeful smile deepened as he leaned closer, his voice soft but filled with curiosity. "If that's where the conversation will lead us."

"Well, I'll think about it. Then give you an answer when you return with the water," Bella quipped.

"I'll be right back," Jack said coyly.

Jack composed himself and returned soon after with two elegant crystal glasses of Gerolsteiner Sparkling Mineral Water. It was all he could do to confidently walk across the room. As he approached the table she was surrounded by a group of men.

Her answer came quickly, first through her eyes, then her smile, and finally with the statement, "Gentlemen, you'll have to excuse us we have a commitment to explore the reality of wormholes."

She elegantly smiled and took the glass from Jack. She thanked him, took a sip and said "Are you all right?"

"I couldn't be better; I know just the place for dinner."

"Was there ever a doubt?" Taking his arm, Bella allowed herself to return to a serene moment, ready to embrace whatever the evening—and Jack—might bring.

CHAPTER 35
HUMAN ENTANGLEMENT

Vieille Ville, Old Town, Geneva, Switzerland
04 August 19:05 Local Time

Two scientists stepped out of the conference hotel into the cobblestone streets of Geneva, their hands brushing together, slipping naturally into one another's grasp. Neither spoke of their work, each other's work, or any area of scientific endeavor. Alchemy, a moment of pure chemistry, unspoken but deeply felt.

Their hands met naturally, slipped together as though it was second nature. They were comfortable in their shared orbit, and as they walked side by side down a cobblestone street, Bella reveled in the cool, summer evening air.

The streets were lined with quaint boutiques and warmly lit cafes, casting a romantic glow over their reunion, and while they chatted and laughed, sharing stories, catching up on the years missed, each moment drew them closer, reigniting an old spark. They approached the lakeside, the sound of lapping water mixing with distant laughter and the soft strumming of a guitar from a nearby busker.

They passed L'Horloge Fleurie, The Flower Clock, one of the largest of its kind in the world. Bordering the Old Town, near the Jardin Anglais, The English Garden. This landmark was first created in 1955 and has since been a symbol of the city's creative and innovative spirit, symbolizing Geneva's watch industry with a combination of floral artistry with timekeeping. The clock face was created with thousands of flowering plants that were changed seasonally. Bella and Jack marveled at the showpiece, neither speaking about its mechanical precision or complexity of the clock's workings. They only saw the beauty of the setting, the fragrance of the flowers, and the artistry of the setting. Chemistry.

Admiring the beauty of the landscape, Bella leaned into Jack's arm, her voice thoughtful. "There is something fascinating with time and the way the Swiss have turned something so common into something of such beauty. Einstein was at the Swiss patent office when he published relativity. What could be more reflective of time than relativity? Only the Swiss."

"You're right, the Swiss make time beautiful. Speaking of beautiful timepieces, I noticed your Rolex Cellini". Jack then lifted his wrist to show his own timepiece. "I enjoy my Milgauss, made specifically for those of us scientists who work with waves, particles, and magnetics. It was a gift from my father." Bella was quiet momentarily, leaving Jack a little confused. She pulled his arm around her and set her head on his chest.

Just beyond The Flower Clock, the old section of Geneva, commonly known as Vieille Ville (Old Town), steeped in history and culture-a charming and picturesque area. The historic heart of Geneva, it was nestled on a hill beside the river Rhône, offering a delightful blend of narrow cobblestone streets, quaint squares, hidden courtyards, and spectacular views.

Their destination was an elegant restaurant overlooking Lake Geneva. Lights from the city danced on the water like skipping stones, setting the perfect atmosphere for their reunion. As they settled into their seats, a shared glance acknowledging the special atmosphere between them.

CHAPTER 36
DAY TWO: LESSONS IN CHEMISTRY

International Physics Association Annual Conference
Geneva, Switzerland, 05 August

Speakers, vendors, and organizers buzzed around the conference halls like bees around flowers, creating an air of palpable excitement as participants engaged with ideas and research. The vendors brought an energy unlike that of any other conference in Geneva, distinguishing this event from the city's annual watchmakers and automobile shows. For a conference like this, its participants seemed like anyone, from any industry, until they were stopped and observed. The displays ranged from technical books and hardware to whimsical items like keychains shaped as unseen particles—electrons, photons, and even a Higgs boson. CERN had a booth. An intellectual step above the day-to-day conference.

"Bella, I saved you a seat!" Jenni called out, waving her hand as Bella entered the keynote luncheon. "I was hoping to see you, but you never know, the way you left the cocktail party with the American."

"I can't really explain it," Bella said. "Something happened. Something I don't really understand."

"What happened is you met someone too cute to be a physicist. Tell me the details, did you take him back to your room, or did you go to his room?"

Bella pinched her tongue between her teeth and shook her head. "You should write romance novels. We didn't end up in either room. He went to his room I went to mine."

"I think the rest of us needed Personal Protective Equipment (PPE) to survive the energy you two created."

Bella filled Jenni in on the details. They'd met fifteen years ago on the Camino when Bella was only fourteen and working for her mother and Bea at the cafe. It was the first time she'd felt Cupid's Arrow, the lightning bolt of love at first sight. "We're going to meet at this evening's reception. I know it sounds a bit Jane Austen-ish, but there feels like there's something to this chemistry."

Jenni teased out the rest of the night's details over the course of the presentation, which neither of them paid any real attention to. The speaker was boisterous, demonstrative and controversial. He solicited challenging disagreement and the audience responded with jeers. After taking and vehemently debating a couple of questions challenging commercialization and capitalism, he was whisked away by security. Jenni and Bella were unfazed, the two were engrossed in the possibilities of romance.

A hotel page arrived at the table with a hotel blue envelope with Bella's name on it. He handed it to her, smiled and departed. How did he know it was her? Bella opened the blue envelope cautiously, her heart skipping as she read the card inside:

Quantum

"Forget the reception, dinner at 19:00. See you in the lobby."

Bella smiled as she embraced the unexpected. Already imagining the evening ahead. Jenni leaned in eagerly, her wicked smirk matching Bella's blush. 'Well, well. That's straight out of a romance novel.' Bella rolled her eyes but handed over the card, if only to stop Jenni from making a scene. Jenni smiled a wicked smirk. She then just nodded.

That afternoon, Bella with a school-girl's anticipation, skipped the afternoon speaker sessions and, realizing she had only business attire, she spent the afternoon browsing the shops of the fashion district. The alchemy was really taking effect, she wanted to be perfect for the dinner date.

Enjoying her time shopping and basking in the anticipation, she selected a cobalt blue, form-fitting flowing pantsuit with complimentary Christian Louboutin pumps. Deciding to wear her hair down, she felt like a teenager attending her prom.

She turned her attention to jewelry. She realized, attending a professional conference, she had packed lightly bringing a few key pieces. She always traveled with a pearl necklace her uncle had created, adding a pearl for every birthday, Christmas, and milestone in her life. She went to the safe to take out her go-to daily wearer Rolex Cellini Moon Phase watch, another gift from Uncle Andrew.

In fussing with her wardrobe, she came across the Sky Dweller meteorite dial. Why she brought it, she had no idea. She had only worn it selectively and never with her co-workers. As she slipped the Sky-Dweller onto her wrist, a strange sense of power and nostalgia washed over her. Why she brought it remained a mystery, but tonight, it felt right. Looking at herself in the mirror, however, Bella felt brand new. When the time came to head out, she approached the elevator down the hallway. As it opened, the gentlemen inside, clearly enraptured in their banter, clammed up at the sight of her. As the elevator descended, Bella caught the shifting glances of the men inside. Their chatter faded into silence, their companions squeezing their hands in subtle acknowledgment of her presence. Bella couldn't help but smile at the unspoken power she seemed to wield. Though she rarely indulged in gloating, the sudden shift in the elevator's atmosphere amused her.

Once the elevator reached the lobby, she walked through to the reception area, spotting the keynote speaker from earlier that day exchanging verbal abuse with a group of scientists. She also saw Jenni, who beamed and fanned herself as if cooling off from a sudden burst of heat.

Just then she felt an arm slide around her waist. She was ready with a quip—ask before you touch someone—but then she saw Jack smile. He whispered in her ear, "You look incredible. Are you sure you're not here to model instead of present quantum theories?"

She gave him a peck on the cheek, leaned into him and whispered. "If it's a dream I'm sharing it with you." Then gave him another kiss on the cheek and surprised him with a slow lip kiss. Turning, she gestured to Jenni. "This is my good friend and fellow physicist Jenni. And Jenni, just so you know- no passes, he's taken!" Jenni smiled and accepted a kiss on both cheeks from Jack.

"I've made reservations at Cafe de Paris and have a car waiting," Jack told them. Jenni gestured to the door as if to hurry them. Bella stared at Jack's hazel emerald green eyes and the lessons in chemistry continued. Smiling, she took Jack's arm and almost led him through the lobby.

A sleek black Mercedes S-Class waited at the curb, its polished exterior gleaming under the evening lights, while a driver held the door open with practiced grace. "You sure know how to make an impression on a girl," Bella said.

"Are you kidding I can take lessons from you on making an impression. You are even more beautiful than I could have imagined."

In the back seat, Bella noticed she was having an impact on Jack. His pants rose and tightened across his lap. As the car glided through the Geneva streets, the soft hum of the engine filled the air. Bella leaned into Jack's arm, inhaling the faint scent of his cologne, a blend of cedar and spice, as their chemistry sparked in the confined space.

Situated on Rue du Mont-Blanc, Café de Paris lived up to its reputation, offering a classic bistro setting with warmly decorated interiors, exquisite mirrors, and wooden accents that evoked timeless Parisian charm.

Bella sipped her Belvedere Cosmopolitan while Jack nursed his Woodford Old Fashioned, their smiles growing wider as their eyes locked. The years melted away, their connection as strong as ever, just smiling and staring into one another's eyes. Those same eyes that connected them years ago. Their hands found one another's as if by instinct, touching and clasping as they shared experiences and dreams.

Dinner could wait, but they held the waiter off as long as possible. The highlight of Café de Paris was their Entrecôte Café de Paris—a perfectly cooked steak served with a secret butter-based sauce, a closely guarded recipe since 1930, golden fries, and a seasonal salad. It was accompanied by a Chapoutier Hermitage, a Rhône wine renowned for its depth and complexity, with layers of blackberry, spice, leather, and earthy undertones.

Their meals arrived, almost too quickly for their liking. As they shared their meal, time seemed to stand still. Dessert was never considered; their thoughts were already entwined elsewhere.

The return drive to the hotel was slow as the streets were packed with crowds. The beautiful day brought out both locals and tourists congesting the route. The pair was more focused on one another rather than on what was going on outside. Bella relaxed across the seat, and settled into Jack's arms. Instinctively, she pressed her thighs together to try to stem the rising excitement. She brought Jack's lips to hers. The kiss was electric, sending current throughout both of them. Jack's tongue met hers, a spark igniting as shivers coursed down her spine.

As they arrived back at the hotel, the crowd waiting at the valet stand or those waiting for local transportation all turned as the driver opened the back door for Bella. She emerged, exuded elegance, her cobalt blue outfit flowing around her like a second skin, her radiant confidence captivating everyone in the lobby. A magnetic presence. All time stopped. Everything else ceased to exist. She almost flowed into the lobby with beauty and grace. Jack followed, smiled with pride and anticipation.

Bella was at her peak. The flowing gauziness of her suit highlighted her curves, her hair cascaded over her shoulders, and her smile beamed. She was a modern Aphrodite, the goddess of desire, lust, beauty, pleasure, passion, and love.

At the elevator doors, Bella entered quickly, closing the doors before any others entered. Her magic continued as the doors silently slid shut. Jack hit the floor button. The car rose. As Bella leaned into Jack, she whispered, 'I've waited too long for this.' Her lips found his, and their shared passion drowned out the world around them. Bella hit the stop button. The elevator suddenly ceased its upward trajectory. But not Jack's. Bella's passion took over. She drew Jack to her. Her lips grazed his neck, trailing upward to his ear before capturing his mouth with a deliberate, lingering kiss. Her leg curled around him, drawing him closer as their passion ignited. He couldn't resist as she drew him in. He pressed his pelvis against her with desire. Jack had never come as close to climax as when she responded with her own thrusts, moaning softly in his ear. Bella was rocketed into her own eruption of passion, she was rocked with quivers. Knowing the certainty of the verge of the mutual eruption of passions before them, Bella wanted to savor their passion in her private domain. Uncertain on her feet she held Jack for stability. She whispered to him to hit the stop button again, setting the elevator in motion. As the elevator resumed its climb, their breaths mingled, and their shared desire promised that the night was just beginning.

CHAPTER 37
BITTERSWEET DEPARTURE

Geneva (GVA) Airport, Geneva, Switzerland
07 August 09:21 Local Time

Their whirlwind days in Geneva had been serendipitous—a chance to rekindle a connection that had endured through years and miles. That spark, first ignited on the Camino fifteen years ago, remained as powerful as ever when their eyes met again in Geneva. They had been youthful yet determined to make their lives meaningful. It was love at first sight, the elemental catalyst that sparked a chemical reaction—one that would never diminish. It was as if they had never parted. Chemistry.

In the evenings, they dined at the best offerings of Geneva overlooking the shimmering Lake Geneva, sharing stories under the night's stars—each moment weaving them closer together. Over dinners, they shared laughter and memories, weaving a bond that grew stronger with each passing moment.

They spent hours in amorous seclusion together, hours of explosive sexual experimentation. Each explored the other's curves, smoothness, hardness, moisture, pheromones, and sweat. The climaxes and releases were like no other either had experienced before. They savored the intimacy, they were natural together.

Even with their knowledge and ability to apply the theories of relativity and quantum, they could not stop time, and it finally became time to return to their research, studies, and homes. At Geneva Airport, their farewell was bittersweet. Amid bustling travelers and overhead announcements, they stood close, reluctant to part.

They made a promise to continue what they had rekindled: he would fly to Oxford; she to Arizona after her return trip to Switzerland and CERN. The farewell was punctuated with a promise to bridge the distance with love and shared intellectual passion.

Now, at the departure gates, the reality of their separate lives settled heavily over them. As they stood close, Bella felt a pang of regret mixed with hope. This was not an end, she reminded herself, but a beginning. He was bound for Arizona, back to his lab where his work as a quantum physicist was entering a crucial phase. She to Oxford and CERN and to her research in quantum networks. They stood, hands clasped, each wearing a smile tinged with sadness. She with a small tear under her eye, he unable to see the tear because he had her wrapped in his arms, her head nuzzled in his neck.

Then her final boarding call echoed through the terminal, shattering their last lingering embrace. With a deep breath, she leaned in for a final kiss, lingering as if to capture the moment forever. When they pulled apart, the weight of their promises hung in the air between them. As he walked toward his gate, he turned back to see her waving, a figure of resilience and grace. They both knew that this was not just a farewell but the start of a new chapter.

The echo of their promises filled their thoughts as they headed home, each carrying a heart full of hope and a mind buzzing with plans for the future. They carried with them the certainty that their paths would cross again—guided by love, purpose, and the unbreakable chemistry that bound them. Geneva had been a beautiful chapter in their story—a prelude to the chapters they would write together, no matter the distance.

CHAPTER 38
JACK AT UA

University of Arizona, Tucson, Arizona
15 August 18:10 Local Time

In one of Jack's early meetings on a return to UA following the conference in Geneva, he met with his mentor. Rick welcomed Jack back from Geneva with a firm handshake, their favorite dinner spot bustling with the evening's energy. It was a busy night at the University Club above the football stadium on campus.

Rick welcomed Jack back from the conference in Geneva. He asked about his time there. Specifically, the conference and the spectacular Genevan views.

"Really nice," Jack answered quickly, struggling to maintain eye contact with Rick. "Really nice."

Rick, ever the relentless mentor, raised an eyebrow at Jack's vague responses, his curiosity unyielding. "Nice, hah?"

When Jack nodded, offering no further explanation, Rick pursed his lips. After half an hour of persistence, Rick finally coaxed enough details from Jack to piece the story together, including the fifteen-year connection to the woman he had spent the conference with. Rick looked thoroughly drained, as if he'd just endured a tedious and exhausting task.

"Dr. Arensico," he said. "Dr. Bella Arensico, from Oxford. One of the most respected researchers in quantum networking and communications in the world. I can't wait to hear about how you two can perfect spooky entanglement at a distance."

"Don't laugh," Jack leaned forward, his voice lowering. "When we were apart, I could still feel her presence, as if some invisible thread connected us. It's not science, but it feels real. I think emotional entanglement is real—something beyond what science, even quantum mechanics, can explain. We sense or feel one another halfway around the globe."

Jack paused, his thoughts momentarily drifting to Geneva, to Bella, and the inexplicable connection they shared. Feeling her presence. Shaking himself free of the memory, he returned to the present. "We did exchange ideas on our research," he continued, his tone steadying, "and I have a new direction I'm going to explore."

"You sure you haven't explored enough for a little while?"

Jack crunched his face up. "I'm serious," he said.

Rick raised his hand in apology. "Okay, I'll refrain."

"I had an epiphany moment," Jack continued. He flicked through the menu, ordering something quickly, and waiting impatiently for Rick to follow suit with the suspicion that he was deliberately taking his time. When he finally made his order, he pressed his fingers together and gave a teasing smile.

"In contemplating quantum networks, the computers themselves become the limiting factor. Everyone from DARPA (US Defense Advanced Research Projects Agency) to Apple are working on the computers. Intel and the chip makers are researching the next generations of semiconductors. It's Intel's work with Josephson junctions as superconductors that is promising but the materials are the limiting factor. I'm thinking there's potential in Josephson junction theory and our material science projects."

A Josephson junction is a tiny structure used in advanced electronics. It consists of a very thin layer of material that does not conduct electricity, placed between two layers of materials that are superconductors (materials that allow electricity to flow without resistance at very low temperatures).

What's special about this structure is that pairs of electrons in the superconductors can pass directly through the non-conducting layer as if it weren't there, thanks to a phenomenon called "tunneling." This makes Josephson junctions useful for creating highly sensitive devices and circuits.

To understand Josephson junctions, it's important to first grasp the basic principles of superconductivity. Superconductivity occurs when certain metals and alloys are cooled to extremely low temperatures—within 20 degrees or less of absolute zero. At this "critical temperature," these materials undergo a phase transition, shifting from a normal state with electrical resistance to a superconducting state where resistance essentially disappears.

In the superconducting state, electrons within the metal exhibit an unusual behavior. Above the critical temperature, the interaction between two electrons is naturally repulsive. However, below the critical temperature, this interaction becomes very slightly attractive. This attraction is due to the interaction of electrons with the ionic lattice structure of the metal. The result? An environment with no electrical resistance, meaning no energy loss during the flow of direct electrical current.

This Josephson junction concept is where Jack is expecting a breakthrough, although not at the supercooling level as that would be impractical for normal use.

"What do you think?" Jack asked. Rick drummed his fingers on the table. As Jack had been speaking, he had seen Rick's face descend somewhat, which Jack initially hoped had been because his idea had buried his mentor deep in thought, but the longer Rick held this posture, the more concerned Jack became. He egged Rick on for a response. His mentor exhaled loudly.

"Rick's gaze flicked toward the balcony that surrounded the dining room, his expression unreadable. 'Don't look now,' he muttered. Jack's stomach tightened as he resisted the urge to turn.

Jack's stomach dropped. "Hmm."

"I think we should head to the lab."

"You see them too?"

Rick nodded, his eyes flashing to one particular place. The balcony. Two men stood with their backs to the partition window. Neither of them had turned to speak to the other, with only one occasionally raising a drink to his mouth.

"They were sitting close before, weren't they?" Jack asked. Again, Rick nodded. At that, they casually finished their dinner, trying at small talk until their plates were reasonably cleared. Jack gave a quick look toward the balcony. They had moved away from the window and were now visibly talking among themselves.

"Are you sure we're not just paranoid?" he asked.

"Just make sure you aren't entangled with them," Rick said.

As they left the restaurant, Jack glanced back. The men were no longer on the balcony. His pulse quickened as Rick clapped him on the shoulder. "Let's get to the lab" Rick said, his tone light but his eyes wary.

CHAPTER 39
BELLA AT CERN

Oxford, UK/ Arizona, USA/ Geneva, Switzerland
CERN Headquarters, Quantum Research Laboratory
Fall Season

Enduring their separation, Bella and Jack tried to bridge the distance through chats, texts, and video calls. Despite their busy lives on separate continents, Bella and Jack found ways to nurture their connection, weaving moments of intimacy into the fabric of their demanding careers. Each month, Bella or Jack crossed the Atlantic to reconnect in person, fulfilling both emotional and physical needs. The long flights and busy schedules were challenging but the time spent together was irreplaceable. Bella's visits to Jack's home in the States were filled with hikes in the cool pine-forested mountains, quiet dinners, and late-night conversations under the stars around a firepit. The sounds of insects, birds, and rustling wildlife blended with the earthy scent of pine and granite, subtly shaped by moisture and sunlight. Jack's home was built on a granite overlook. In the evening, the stone would have a slight earthy or mineral-like smell, a sensory counterpoint to the sterile lab environment. Bella relished the simplicity and tranquility.

When Jack traveled to Europe, he immersed himself in Bella's world of particle physics. He was awed by the historic charm of Oxford and Bella's descriptions of the cutting-edge innovation of CERN, the marriage of the old and new. Together, they explored the quaint streets of Oxford, enjoyed meals in cozy pubs, and shared quiet moments in Bella's favorite spots. Highlights included family dinners at Bentley Hall, followed by cozy evenings at Bella's charming home on Magpie Lane.

Each visit became a grounding force, a reminder that while their research was groundbreaking, their love gave their lives purpose and meaning. The physical presence of one another provided a grounding force, reminding them of the life they were building together despite the miles, the continents, and the demanding nature of their careers.

During one of Bella's visits, she and Jack walked a secluded trail, the forest's rhythm providing a soothing backdrop. "I miss this," Bella said softly. "The simplicity of being with you."

Jack smiled, tightening his arm around her. "I miss this, too."

In those shared moments, the frustrations and challenges of their respective research melted away. They found strength in each other's presence, a reminder that their work, while important, was only one part of their lives. Each visit rejuvenated their spirits, providing the emotional fuel they needed to return to their groundbreaking work with renewed vigor.

With each goodbye, they returned to their work renewed, their partnership as dynamic and unshakable as the quantum particles that had brought them together. And with each reunion, they reaffirmed their commitment not just to their work, but to each other, forging a partnership that was as resilient and dynamic as the particles they sought to understand.

Bella's research at CERN was finally within reach. She had worked tirelessly to secure her slot at CERN, and now the moment had arrived. The Large Hadron Collider promised answers to questions few dared to ask. The anticipation kept her focused and driven—the opportunity to work with the collider promised insights into the subatomic world that few could imagine. Her interactions were brief but intense, exchanged with equally driven colleagues. Conversations were terse, filled with technical jargon and the latest findings.

Meanwhile, Jack immersed himself in his work, finding solace in his daily routines. Yet, Bella's absence had been a constant shadow, reminding him of the moments they had shared and the promises of what lay ahead.

Bella focused her mind on her work. Scientists from around the world converged here, creating a sprawling complex buzzing with activity. Bella felt a familiar thrill as she donned her lab coat and made her way to the control room. The collider, a marvel of modern engineering, represented the pinnacle of human curiosity and ambition.

She delved into the data, seeking patterns and anomalies that could unlock new understandings of the universe, and as the hours melted away, she found herself lost in the intricate dance of particles and waves. Yet, in the quiet moments, her thoughts would drift to Jack. She remembered his touch, the way his presence could ground her even in the most chaotic of times. The texts and calls were not enough. She longed for the warmth of his embrace, the simplicity of their time together.

Quantum

As days turned into weeks, Bella's research yielded promising results. She felt the thrill of discovery, the sense of being on the brink of something monumental. But the joy was tinged with a lingering ache. She knew that Jack was waiting, just as eager for their reunion as she was.

Their separations had only intensified their bond. Bella resolved to finish her work with the collider and return to Jack as soon as possible. The complexity of quantum networks and the simplicity of their love were two sides of the same coin, each fueling her passion and determination.

With her research at a critical juncture, Bella allowed herself a moment of reflection. She remembered Jack's Rolex Milgauss, an object of beauty and precision, much like the man himself. Jack embodied a complexity and simplicity that Bella found endlessly fascinating. He could discuss the intricacies of quantum mechanics with her and then switch to a serene contemplation of nature without missing a beat. And for Bella, that was enough. Bella smiled at the memory of their last conversation, in which Jack had compared the steady ticking of his watch to the rhythm of their lives. "In a world of chaos, there's beauty in the predictable," he had said, his eyes twinkling with warmth. The lab's clock struck midnight, pulling Bella back to the present. She stretched, feeling the fatigue in her bones. She glanced at her screen, where streams of data awaited her analysis. She thrived on the complexity, just as Jack thrived on simplicity.

Hours later, as dawn approached, Bella finally allowed herself a brief rest. She closed her eyes, the image of Jack and his comforting smile lingering in her mind. The sterile hum of the lab was a sharp contrast to Jack's warm embrace, a memory Bella clung to in her moments of solitude. In the labyrinth of quantum physics,

Jack was her constant, her anchor. For now, Bella returned to her research, driven by the dual forces of scientific curiosity and the quiet, steady love that kept her grounded. Their late-night, early morning or mid-day conversations were filled with discussions of quarks, entanglement, and the latest breakthroughs in particle physics. Jack's updates often served as a balm to Bella's frustrations whenever she came across obstacles in the research, his successes igniting a renewed determination within her.

Bella knew that Jack's advancements would provide the missing pieces she needed. The theoretical models he was developing offered new perspectives that could be applied to her quantum network research. Conversely, the foundational work Bella was conducting on particle behavior could offer Jack the empirical data required to validate his theories. As weeks turned into months, Bella's perseverance began to pay off. Small but significant breakthroughs started to appear in her data, hinting at the larger picture she was trying to piece together. Each step forward, however incremental, felt like a victory. She found encouragement in the knowledge that Jack's achievements were, in a way, paving the path for her own. One evening, while analyzing a particularly perplexing set of results, Bella had a sudden insight. She realized that one of Jack's recent discoveries about materials and particle interaction could be the key to unlocking her own research's potential. Excited, she immediately called Jack, her frustration momentarily forgotten in the thrill of collaboration.

“Jack, I think I've found something," she began, her voice trembling with excitement. "Your latest paper on crystalline structures related to quantum coherence—it might be the piece I've been missing.”

Bella’s voice trembled with excitement as she explained her discovery, her words tumbling out in a rush. Jack listened, his own enthusiasm building as he pulled up her data on his screen. The data was promising. They spent the next few hours brainstorming and mapping out the next steps. In that moment, the physical distance between them seemed to disappear, replaced by the electric connection of shared intellectual pursuit.

Their work, once parallel, was beginning to converge. Bella's frustration was giving way to a renewed sense of purpose, fueled by the knowledge that together, they were inching closer to a groundbreaking discovery. The complexity of their research was balanced by the simplicity of their bond, a partnership that thrived on mutual respect and the unyielding quest for knowledge.

As they signed off, Bella felt a calm she hadn't known in weeks. Their shared purpose was rekindling her spirit, reminding her why they both chased answers in a world of uncertainty. She knew there were still many challenges ahead, but with Jack by her side—if not physically, then in spirit—she felt ready to face them.

CHAPTER 40
WHAT'S HAPPENING

Research Lab of Optical Science Building University of Arizona, Tucson, Arizona
30 October 13:37 Local Time

In the solitude of his basement lab, Jack conducted his research in advancing material sciences. The materials science and quantum exploration lab at the University of Arizona was a cutting-edge facility, advancing material sciences. The lab housed advanced equipment, including cryostats for studying materials at extremely low temperatures, scanning tunneling microscopes for imaging surfaces at the atomic level, and various spectrometers for analyzing material properties.

In the quiet, controlled environment of the basement lab at the University of Arizona, Jack delved into the mysteries of material science. Located three stories underground, the lab was a fortress of innovation, shielded from environmental interference and designed for precision experiments. Entering the lab required passing through multiple layers of security. 'Badging in' was a sophisticated process that involved scanning an ID badge, facial recognition for identity verification, and inputting a unique key code. These measures ensured that only authorized personnel could access the secure facility, located deep within the basement of the Optical Sciences building. Inside, researchers—including PhDs, post-grads, and graduate assistants in lab coats—were engrossed in their experiments. Each person worked diligently on advancing quantum computing elements, superconducting materials, or innovative semiconductor devices.

The atmosphere in the lab over the past few weeks had been electric. Led by Jack, a team of scientists had been working on a highly ambitious project: to engineer a crystalline material with superconducting properties at temperatures higher than ever before achieved while applying the Josephson junction theory. Using a process known as 'doping,' they replaced atoms in the crystalline lattice with other elements, altering the material's electronic properties.

Jack and the team were exploring advanced crystalline structures while applying the Josephson junction theory. The current problem with superconductors is that they must achieve critically low temperatures to experience their performance. To solve this problem, they conducted a delicate operation where individual atoms were replaced with surgical precision, using advanced tools including a scanning tunneling microscope capable of manipulating single atoms. Throughout the experiment, the team monitored the changes in electrical resistance and other key properties that indicate the emergence of superconductivity.

Meticulously making final adjustments and applying the laser technique that produces the Josephson junction, the critical moment arrived. As the team gradually raised the material's temperature from near absolute zero, a hushed tension filled the lab. Each data point brought them closer to an answer, and every eye was fixed on the monitors. Time seemed to freeze as the minutes turned into hours. Jack glanced at his Milgauss repeatedly, reassuring himself that time had not truly stopped.

Then, celebration erupted. The material had maintained zero electrical resistance at a record-breaking room temperature. They had not only created a new superconductor but shattered the ceiling for operational temperature, bringing the dream of room-temperature superconductivity to reality.

Amid the excitement, Jack called for calm as they initiated a series of rigorous verification tests. The material was subjected to magnetic fields, electrical currents, and various stress tests to confirm its superconducting properties. The data came in, affirming their discovery. There was no electrical resistance in the superconductor, and therefore, no energy loss.

Finally, the empirical proof, a breakthrough equivalent to Alexander Graham Bell's famous "Watson, come here, I want to see you" moment. A monumental milestone in quantum computing, physics, material science engineering, and information science. A significant leap forward in quantum computation. The holy grail of quantum computing.

Verifying the results, the mood shifted from tense to triumphant. The team shared handshakes and embraces before Jack quickly ordered them to document every step of their experiment, taking meticulous notes and photographs, and securing the samples. Their findings had to withstand the scrutiny of peer review and replication by others in the field.

They gathered around a breakroom table laden with pizza and Chinese carryout boxes, snacks and drinks. The once-tense atmosphere transformed into joyous chaos. Handshakes turned into embraces, and a long-chilled bottle of champagne was finally opened, poured into beakers that stood in for flutes. Their discovery held transformative potential: lossless power grids, frictionless maglev trains, and quantum computers with unprecedented efficiency. The possibilities were as vast as the challenges ahead.

There was an urgency to share the news with the scientific community and the public. However, Jack reminded the team of the weight of their discovery, cautioning them about confidentiality. For now, the world would remain unaware of the breakthrough poised to reshape technology and science. A reminder they were bound by confidentiality agreements. They could not discuss this work outside of the lab. The lawyers, funding agency, and university leadership would control the dissemination of any information. Nevertheless, there was excitement about the future as they stood on the precipice of a new technological era, one that they ushered in.

CHAPTER 41
THE DISCOVERY

Research Lab of Optical Science Building University of Arizona, Tucson, Arizona

30 October 23:37 Local Time

That evening, Jack stayed behind in the lab while the others headed home or out to celebrate. The hushed environment of the basement, punctuated by the soft hum of supercomputers and the whir of cooling systems, provided a strange comfort, steadying his nerves after the day's breakthroughs.

During a celebratory call with Bella, Jack had to hold back from sharing all his good news. "I'll tell you everything in two days when I visit Oxford," he promised. They raised virtual glasses—Bella with her favorite wine and Jack with the last of the champagne from earlier that day. Bella knew he had achieved something significant and was eager to hear more. The anticipation of waiting to hold one another, to share their work and lives in person, hummed like an invisible energy field.

After ending the call, Jack leaned back in his chair, letting the implications of his discovery sink in. The soft hum of lab equipment faded into the background as he processed the enormity of it all. He had stumbled upon something that went beyond their original research objectives—a breakthrough that could change the landscape of energy production. He wanted to savor his accomplishment. He stared at the screen, the numbers and graphs flashing before him, each one a testament to the uncharted territory he'd just crossed.

He couldn't shake the realization that this wasn't just a minor anomaly or a fluke. What they'd achieved earlier was a paradigm shift. A combination of the tritium-based crystalline lattice and Zeliconite had triggered a controlled neutron modulation that, on a fundamental level, shouldn't have been possible outside massive, high-temperature fusion reactors. Yet here it was—a stable, contained neutron modulation fusion reaction on a tabletop scale.

The neutron imbalance was the key—a subtle imbalance that allowed the reaction to occur without spiraling into a destructive cascade. Unlike the runaway chain reactions of nuclear fission, this reaction was calm and controlled, resembling the steady burn of a star. The Zeliconite, which had initially been intended as a simple support material, was acting as both a generator and a modulator of neutrons. It was the missing piece they hadn't even known they were looking for.

This hinted at deuterium-tritium fusion (DTF), a powerful reaction long considered the holy grail of clean energy. The process typically required extreme temperatures and produced helium-4, a free neutron, and a 17.6 MeV release of energy. This is a very small amount of energy when viewed on a macroscopic scale, but in nuclear physics, it represents a significant amount of energy per reaction due to the enormous number of particles involved in practical applications like fusion. Yet Jack's discovery had achieved this at a level of stability and efficiency that bordered on miraculous.

Jack's heart raced. This wasn't just a scientific breakthrough—it was the dawn of a new era in energy production. The implications were staggering: lossless power grids, limitless energy for quantum computing, and a solution to humanity's dependence on fossil fuels. He glanced at the clock. It was well past midnight, but his mind refused to rest. He considered rushing to Dr. Stephens' quarters to share the news but held himself back. Verification was crucial. They'd need to replicate the results, run safety tests, and analyze every variable.

They'd spent years chasing small gains, painstakingly adjusting variables and recalculating hypotheses. And now, this.

This discovery demanded careful consideration. They would need to verify the results, run additional simulations, and find ways to stabilize the process even further. And there were safety concerns, too; they were dealing with fusion on a scale that, if mishandled, could have unintended consequences. The neutrons themselves could become a hazard if not properly contained. Jack knew that cold fusion, while safer, was still unpredictable. Every variable, every change could shift the balance, so they'd have to tread carefully.

Cold fusion is a theoretical process where nuclear fusion—combining atomic nuclei to release energy—occurs at or near room temperature. This contrasts with traditional nuclear fusion, which requires extremely high temperatures, like those found in the sun.

If achieved, cold fusion could provide a nearly limitless, clean energy source without the harmful radiation or environmental impact of current atomic energy technologies. However, it remains a controversial and largely unproven scientific concept.

He took a deep breath, his mind already churning with ideas for the morning. The first thing he'd do would be to gather Bella and Dr. Stephens for a thorough review. They'd go over every piece of data, every calculation, until they were certain. And once they had confirmed the results, they could begin exploring applications, figuring out ways to harness this newfound power without compromising safety.

For tonight, he'd keep the discovery to himself and his team. He needed the quiet of the night to process what he'd uncovered and prepare himself for the conversation ahead. Tomorrow would mark the beginning of a journey that promised to redefine everything they thought they knew about their work and the possibilities of fusion energy.

As he saved his notes and powered down the equipment, a faint unease prickled at the back of his mind. The normal hum, underlying the silence in the lab, felt heavier than usual, pressing against his senses. Brushing it off as exhaustion, he packed up his materials and prepared to head to his quarters. Jack dimmed the lights in the lab. As he made his way out on his way to his quarters, his thoughts drifted to the broader implications of their work.

With one last look back at the dim glow of his workspace, he whispered to himself, "This is only the beginning."

Then he heard it—a faint scraping sound, like metal against concrete. Jack froze, his hand hovering over the lab table. An unmistakable scrape echoed again, this time closer, like the faint movement of an object. His pulse quickened as he reached for his phone and opened the lab's Nest app to check the security feed. All doors were locked, but the sight of two men standing outside the main door sent a chill racing up his spine.

The main door rattled. It sent a shiver of unease through him. Moments later, the keypad beeped faintly. Someone was trying to access the lab.

Jack's mind raced. On his phone he accessed the security app linked to the lab's door systems. With a few taps, he engaged the deadbolt privacy mode, disabling the keypad. The men on the feed looked frustrated, their faces illuminated by the faint glow of the keypad. Jack's chest tightened as he recognized them—they were the same men he'd seen at the club earlier that week.

Dialing 911, the call was quickly connected to the University of Arizona Police Department. He explained the situation in hushed tones. The dispatcher assured him that officers were on their way and advised him to stay calm and to stay on the line. With his eyes glued to the security feed, Jack waited anxiously. The two men vanished from the screen but whether they had given up or were about to attempt something else was anyone's guess.

The next few minutes felt agonizingly long. The dispatcher asked Jack how he was holding up and informed him that the officers had arrived at the building. Her calm demeanor was reassuring as she explained that the officers would first clear the building for safety before they made their way to the lab. Finally, the sound of footsteps reached his ears—this time, purposeful and steady. The camera feed showed four uniformed officers approaching the lab door. One by one, they displayed their badges to the camera. The dispatcher confirmed their identities, and Jack carefully disengaged the deadbolt to let them in.

The officers introduced themselves, moved quickly, their eyes scanning the room for signs of disturbance. Sergeant Berriman, the lead officer, introduced himself and guided Jack to a seat. "Tell me what happened," he said, his tone calm but firm.

Jack recounted the events, his voice steady despite the adrenaline coursing through him. He described the two men at the club, their unexpected appearance at the lab, and the breakthrough he and his team had achieved earlier that day. Berriman listened intently, his expression shifting from curiosity to concern.

"Did they say anything?" Berriman asked. Jack shook his head. He felt the chair beneath him and tried to relax into it. He thanked Berriman for their speed in getting to the lab, to which the officer calmly nodded.

"Have you had any threatening emails, texts, calls?"

"No."

" Do you have any disgruntled students? Anyone who may want to harm you?"

Jack thought hard. The question, while boilerplate, was bizarre to even hear.

"No."

The other officers gathered around Jack.

"Anything?" Berriman asked them.

"No one else, sir," one of the other officers said. Berriman's brow raised in acknowledgement before he turned back to Jack.

"You said you recognized them?"

"I recognized them, but don't know them. They looked like they were of Middle Eastern descent, fighting-aged men." Jack recounted his observation of the two men at the club.

"Why do you think they may be interested in you or this lab?"

Jack shrugged. "Something related to our research, maybe. Our work isn't exactly classified."

"And you believe that's relevant?"

"I don't think it's a coincidence," Jack said. "There's nothing else that would draw any attention to me in this way." Then he shared an approximation of the significant breakthrough they had achieved earlier in the day at the lab. The results were not just a step forward but a leap in quantum technology, potentially valuable to both private and governmental entities. Given his military background, Jack explained his belief in the principle that there were no coincidences in such matters. The timing of the two men's appearance at the club then showing up at the lab was too precise to be mere happenstance.

Berriman paused for a moment, his face making slight movements as if he were talking to himself in agreement.

"I too don't believe in coincidences." He took out his phone, tapped it a few times, and brought it to his ear. More a directive than a question he said to Jack. "Can you stick around?"

"Of course," Jack replied.

Berriman retreated to make the phone call. He talked sharply for what seemed like seconds on the phone before thanking whoever was on the other end and hung up.

Berriman ordered two officers to watch the perimeter of the building. The other officer was to stay in the building lobby and watch for any activity. The building was in a lock down mode.

Changing focus, he said, "Someone should be here in half an hour. You got coffee here?"

Jack pointed to a half-full coffee pot next to the fridge and Berriman helped himself. He chuckled, police station coffee, it's been on the burner a while. Taking his coffee he invited Jack to show him around, which Jack presumed was done to help calm his nerves. He explained his work to Berriman, quantum for the non-scientist, and his plans to travel to the UK to see his girlfriend, a fellow quantum researcher.

"Jack, does she know about your research? Is she up to speed with your findings today?"

"We had a call but I didn't share much other than just that we had a breakthrough today. I know not to discuss my findings on an open line. There is always the concern for industrial espionage. So, no she's not."

Eventually, two FBI agents arrived and made introductions and displayed impressive gold credentials. Both women, one in her late forties and one in her early thirties.

"I'm Special Agent in Charge Mickey Golden and this is Special Agent Rita Lopez, our national security expert." Jack was taken aback by the young woman being the national security expert. Then again, she wasn't much younger than him, and he had just achieved the greatest scientific breakthrough of the century.

"Can you bring us up to speed?"

Jack laid out his theory to the agents, describing a direct nexus between his work's breakthrough and the sudden interest these individuals showed. Perhaps they were attempting to gather intelligence or even interfere with the project due to its potential implications and value.

The agents listened intently, taking notes, their expressions revealing little. SA Lopez eventually excused herself and walked to a corner of the lab to make a call. SAIC Golden continued to gather information. When Lopez returned, she whispered something to Golden, who nodded gravely.

"We've elevated this to Washington," Golden said. "They've determined this is a matter of national security. For now, we need to seal your lab until we can clear all team members. We acknowledge that this research is yours, and you'll retain access to all your results. But starting with you, we need to clear everyone."

Jack struggled to respond, but before he could speak, Berriman interjected. "If no one on your team has anything to do with this, it'll be a minor nuisance at worst. As I've said before, no coincidences."

"You will also be required to sign a national security classified confidentiality agreement," Golden continued. "We understand you'll be traveling to the UK soon. You'll need to be debriefed before traveling and again upon your return. On a final note, we've identified the two men from the security video. They're students—much older than usual—Iranian nationals. Efforts to locate them are underway. For now, Sergeant Berriman will remain your contact. This might just be the beginning."

With all of the security documents completed, the agents left the lab. Berriman assured Jack that UAPD would increase their surveillance and collaborate with the other security agencies to ensure the safety of the lab and the integrity of the research. He insisted he stay with Jack in the lab while he finished the remaining tasks for the day. Jack completed his work, the two thoroughly secured the lab together, making sure all sensitive data and equipment were safely locked away.

After securing the lab and the building, Berriman insisted he drive Jack to his car. They worked on a plan for the morning. Given the gravity of the situation and the potential implications for the rest of the lab staff, Jack expressed his desire to notify his colleagues about the incident as soon as possible. Berriman, however, advised discretion until the morning to allow the investigative team to gather more information and present a coherent narrative and safety plan to the lab staff collectively. Jack simply nodded in agreement.

They spent the rest of the journey in silence. During the silent drive, Jack reflected on the adrenaline rush of the past few hours, which was now beginning to fade. He wanted the safety of his home now, sleep away the stress and, maybe, find that in the morning this and all been a fever dream.

To ensure Jack's safety extended beyond the university grounds, Berriman followed Jack home, near the University in the historic Sam Hughes neighborhood. The homes were stately, well-groomed and restored to beyond their 1920s and 30's elegance. At the door, they shook hands.

"Thank you, Stg. Berriman," Jack said.

"Call me Joe," the officer replied. "There'll be a patrol in the area tonight. Sleep well."

Before heading to bed, Jack decided to send his team a message to set the tone for the next day:

"Congratulations on the great work! This has been an exciting journey and we're only beginning. Take the morning off, sleep in. Tomorrow morning meet me at 9:00 in the conference room on the first floor. I'll bring breakfast. Go directly there. Don't go to the lab. I'll bring you up to speed on a discovery that occurred last night. Sleep well. See you in in the AM."

CHAPTER 42
THE BRIEFING

Optical Science Building, University of Arizona

Tucson, Arizona

31 October 08:40 Local Time

The next morning, Jack stopped by the bagel shop on his way to campus, picking up breakfast essentials—bagels, cream cheese, yogurt, fruit, and coffee. He was the first to arrive at the conference room, meticulously setting up the spread, using the task to distract himself from the lingering tension of the night before. When Sgt. Berriman and the agents arrived, Jack busied himself pouring coffees, anything to ease the anxiety gnawing at him. As his team members trickled in, he announced that the meeting would begin once everyone was present.

The atmosphere in the conference room was markedly different from the usual hum of activity in the lab. Instead of the comforting whir of machinery and easy camaraderie, there was a palpable tension. The sterile conference room felt foreign, filled with unfamiliar officials, somber expressions, and an air of uncertainty.

Jack, despite the restless night he had endured, stood to open the meeting. His voice was calm but measured as he introduced the FBI agents and Sgt. Joe Berriman. He recounted the events of the previous night—the suspicious individuals at the club, their appearance at the lab's door, and their attempts to gain entry. Though the weight of the ordeal still hung heavy on him, he was relieved to see understanding in the eyes of his colleagues.

After Jack's account, the FBI agents addressed the room, providing updates on the investigation. They reviewed the sensitive nature of the lab's research which potentially explained implications of the breach. Given the gravity of the situation, enhanced security protocols would be enacted immediately. Each staff member would undergo a thorough background investigation, participate in personal interviews, and sign updated national security confidentiality agreements to supplement existing ones.

The announcement was met with unease. The team, accustomed to the collaborative environment of shared governance, began to push back. Questions flew across the room: Why hadn't they been informed sooner? When would they regain access to the lab? Why weren't they given more of a say, after all they felt entitled to shared governance.

The agents maintained their composure, emphasizing that while everyone was considered trusted, the investigation demanded meticulous compliance to ensure the safety of their work and personal security. "This isn't about mistrust," one agent explained. "It's about protecting everyone here and safeguarding the groundbreaking research you're conducting."

Jack stepped in, urging his team to cooperate fully. He acknowledged their frustrations but underscored the importance of the measures. "I trust each of you completely," he said. "But this process is necessary—not just for the lab, but for all of us. Once these steps are complete, the lab will be accessible again by this afternoon."

By the meeting's end, the team appeared informed but understandably strained. As they filed out, Jack took time to speak with each person individually, offering reassurance. The shift from the previous night's celebrations to the sudden lockdown was jarring, but gradually, the team began to accept the new reality.

As the last of the staff left the room, Berriman and Golden lingered behind. Berriman held the team roster in hand and gestured for Jack's attention. "We appear to have everyone accounted for—except Ms. Hala. Has she contacted you this morning?" His voice was even, but there was a hint of concern beneath his professional demeanor.

Jack frowned, his brow furrowing. "No, I haven't heard from her. I noticed her absence, too. So I called her cell as the team dispersed, but it went straight to voicemail." His mind raced with possibilities. Hala was never one to miss a meeting, let alone an emergency briefing of this magnitude.

Golden exchanged a glance with Berriman, her expression tightening. "Did she mention anything yesterday? Any reason she might be absent today?" she asked, her tone probing but not accusatory.

Jack shook his head. "Not that I recall. She seemed fine when we last spoke—focused, as always. But now…" He trailed off, the unease creeping into his voice.

Golden nodded, her gaze steady. "I'll send someone to her home for a welfare check. Let's not jump to conclusions yet. It could be something as simple as a dead phone battery or her running late. Still, given the circumstances, we need to account for her."

Jack nodded, but the explanation didn't sit well with him. Hala was meticulous, almost to a fault. She would have contacted someone if there was a delay—especially now, with so much at stake. A gnawing sense of dread began to settle in his stomach.

Berriman clapped a reassuring hand on Jack's shoulder. "Let's stay focused on the tasks at hand. We'll get this sorted."

Jack managed a nod, but his thoughts lingered on Hala. As he watched Golden make a call, likely dispatching agents for the welfare check, he couldn't shake the image of her empty chair in the meeting room. It was a stark reminder that, even amidst the controlled chaos of the morning, the absence of one could ripple through the entire team.

CHAPTER 43
JACK'S ARRIVAL IN THE UK

London and Oxford, England Heathrow International Airport & Paddington Station
02 November

Jack's journey to London is charged with a mix of excitement and anticipation. After the long flight, he landed at Heathrow Airport, feeling a surge of enthusiasm pulse through him as he stepped onto British soil. With so much news to share with Bella, every moment brought him closer to the woman who had re-entered his life—every day of their separation had only deepened his appreciation for their reconnection.

Navigating through the bustling terminals of Heathrow, Jack's thoughts were filled with images of Bella: her smile, her intellect, and the way she seemed to fit perfectly into the void his life had unknowingly harbored. With his luggage in tow, he made his way to the Heathrow Express, the quickest link between the airport and Paddington Station.

The train, renowned for its efficiency, offered Jack a moment to collect his thoughts and prepare for the reunion. As the cityscape passed him by, Jack reflected on the remarkable circumstances that brought Bella back into his life, the chance encounter at the Quantum Conference in Geneva and the immediate chemistry that seemed to align just right. His heart raced with the thought of rekindling what had once been put on pause.

Jack's introspective thoughts dissolved the moment the train arrived at Paddington Station. Built in 1854 by the celebrated engineer Isambard Kingdom Brunel, Paddington Station stands as one of London's grand rail terminals, a blend of Victorian charm and historical significance. Today, it remained a vibrant gateway, blending its rich history with modern functionality, featuring a vast train shed with a roof made from three spans of wrought iron arches, a pioneering construction method for the time, that allowed for a wide, open space beneath, free from columns, creating an expansive and airy environment.

Inside, Paddington Station had evolved to meet contemporary needs while respecting its heritage. The station boasted the modern-day comforts of shops and cafes seamlessly integrated into its historic framework. Digital displays and signage juxtaposed with classical architecture.

Jack stepped off the express train and moved through the station with a blend of nerves coursing through him from traveling and the anticipation of seeing Bella. The Mad Bishop Pub at Paddington Station came into view. He entered the pub crowded with travelers. As Jack navigated through the crowd, his eyes darted about the place. Bella, already waiting, spotted Jack first. Her face lit up with a broad, eager smile. The moment Jack saw her, his heart leapt. The noise and rush of the station faded into the background as they locked eyes, feeling as if they were the only two people in the station.

Jack dropped his bags once they were within reaching distance. The reunion was marked by a warm, enveloping embrace, the kind that speaks volumes, conveying years of missed opportunities and the joy of reconnection. Their hug was long as they savored the closeness they had fantasized about during their time apart. How good the two felt together.

Pulling back slightly, Jack and Bella exchanged smiles, their expressions a mix of affection and disbelief at being together again after so long. They settled into a secluded spot in the pub, a cozy corner with a view of the station's comings and goings, perfect for people-watching and intimate conversation. As they sat, their hands found each other across the table, fingers intertwining naturally.

They ordered pints of ale—golden, frothy, and refreshingly crisp—along with plates of fish and chips and, of course, mushy peas, the British classic that Jack had been craving. The food, hearty and comforting, was exactly what Jack needed as they eased into their reunion.

As they sipped their ales and enjoyed the crispy batter of the fish, with Jack especially appreciating the peas, their conversation flowed effortlessly as they caught up, steering clear of the heavy topics that linger in the back of their minds. They reminisced about Geneva and expressed how much they had missed each other. Laughter and affectionate glances filled their time together, reinforcing the deep bond they shared despite the years apart since their first encounter.

The decision to keep the conversation light was mutual. Both understood the need to discuss the recent break-in at Jack's lab and the implications of their respective research, but they also recognized the importance of this personal time. The pub, with its public setting and occasional interruptions by other patrons or staff, wasn't the right place for such sensitive discussions. They planned to delve into those topics later in the privacy of the first-class cabin on the train to Oxford where they could speak freely without fear of being overheard. For now, they relish the simplicity and joy of being together again, enjoying the present moment before facing the complexities waiting down the line.

Gathering their belongings, they stepped back into the bustling energy of Paddington Station and moved with the flow of the crowd, their steps synchronized as they navigated through the busy terminal. The announcement of their train came over the loudspeakers cutting through the hubbub of the station and directing passengers to the correct platform.

As they reached their platform, Jack and Bella were greeted by the sight of the sleek, modern train waiting to whisk them away. They boarded the first-class compartment, a sanctuary of comfort with plush seats and ample space. The cabin was quieter and more secluded, a stark contrast to the buzz of the pub and the station. They selected seats with a worktable between two sets of seats. Jack stowed their luggage. Bella wanted to sit next to Jack but resisted the urge, choosing to sit across from him to hold their place. They chose these seats strategically to keep their conversation private while having a view over the entire cabin watching the other passengers. The coach, thankfully, hardly filled—their strategic location was safe, away from any other passengers.

Settling in, Jack invited Bella to sit next to him. She smiled as she joined him in seats next to one another. The train gently rumbled as it began to move. Now, in the privacy of their accommodations, they prepared to delve into the more serious discussions about their research and the recent unsettling events at Jack's lab.

As the train glided smoothly towards Oxford, the shifting landscapes outside the window provided a serene backdrop to Jack and Bella's conversation. Jack's excitement was palpable. He was eager to share the progress his team had made. Shielded from the outside world, he began to explain the innovative approaches and methodologies they had applied, describing how these strategies had pushed the boundaries of current quantum computing material technologies. He described his work with the crystals and reorganizing the atomic structure. Of course, Bella understood the technology but was nonetheless impressed with Jack's innovation. His enthusiasm grew as he spoke about the implications of these advancements, highlighting how they could revolutionize quantum processing and solutions across multiple industries.

Jack chose his words carefully, deliberately omitting the peculiar generation of extra neutrons during the experiments. He specifically held back on mentioning the unexpected generation of extra neutrons during their experiments. This peculiar anomaly was something he wanted Bella to observe firsthand without any preconceived notions that might skew her analysis. He believed in the value of fresh, unbiased insights, especially from someone as skilled and knowledgeable as Bella. Jack knew that her conclusions and interpretations could provide crucial insights into understanding this phenomenon, potentially leading to even greater discoveries.

As he shared his excitement and the broad strokes of the project's success, he watched Bella's reactions closely, her interest and understanding deepening with each detail he provided. The conversation was not just a briefing but an invitation to a deeper partnership in exploring the frontiers of quantum science.

The porter came and provided coffee for Jack, tea for Bella, and a crisp white Pinot Grigio for each along with cheese and biscuits. Careful to ensure privacy, Jack held back until the porter moved on. When they were once again alone, Jack recounted the unsettling events that had transpired at his university lab and the steps taken post-incident—the involvement of the university police, the security measures implemented at the lab, and the ongoing investigation. His tone conveyed a mix of frustration and concern, not just for the security of his work but for the potential implications it could have on national security. Jack didn't want to accidentally put Bella at risk of harm.

As the conversation deepened, Bella listened intently, her own curiosity piqued by the implications of Jack's research and the motivations of the two men. Ever the scientist and pragmatist, she encouraged Jack to consider all possibilities, including the chance that their interest might be academically driven rather than nefarious. As Jack finished sharing the details of his team's breakthrough, Bella quietly processed the information and compared it to her own experiences. She took the opportunity to voice a concern she had been harboring yet had not previously shared.

Bella recounted a troubling incident in her lab: one morning, she found research materials out of place and equipment settings subtly altered. Initially, she thought it might have been an oversight by her team, but as she delved deeper, it became clear that the disruptions were too systematic to be accidental. She also mentioned that security cameras had captured images of individuals, middle-eastern men, who were not recognized as part of the campus community, lurking near her building at odd hours.

These observations had led her to suspect an intrusion, possibly with an intent to tamper with or steal sensitive research data.

When Bella shared her concerns about her lab, emphasized the gravity of their shared experiences to deepen the suspense. The eerie parallels between their incidents left them both uneasy, hinting at a broader, more sinister pattern. The parallels between their experiences were striking and unsettling. Both were working on cutting-edge research that, if misused or exposed prematurely, could have significant implications. There was a distinct concern to be had not only about the safety of her research but also about the broader security implications for the field and potentially national or global security.

Jack listened with growing apprehension, realizing that what he had initially thought to be an isolated incident at his lab might be part of a larger pattern of targeted actions against quantum computing research. The discussion turned towards strategies for bolstering their lab security, the importance of communicating with their respective university security teams, and the potential need to further involve law enforcement agencies.

As the train continued its journey, Jack and Bella began to sketch out a plan for a joint response to their security concerns. Jack would reach out to Joe Berriman and request an alert for his British counterparts at MI5.

As the train announced its arrival at Oxford, Jack and Bella readied themselves. Jack grabbed their luggage. They stepped into the city that now held their next chapter.

CHAPTER 44
THE REUNION

Oxford Station and Bentley Hall Oxfordshire, England
02 November 15:22 Local Time

Jack and Bella's minds remained occupied with the serious discussions they had shared during the journey as they stepped off the train and into the station where a trusted driver and family confidante was already waiting for them. Charles had already carried Jack's luggage to the Bentley estate car, the imperial blue BMW 750 gleaming in the sunlight. Jack and Bella, appreciative of Charles's professionalism and discretion, settled into the plush leather seats of the BMW. Charles greeted them warmly, his familiar presence providing a much-needed sense of security and ease after the intensity of their train ride.

The journey to Bentley Hall proved to be a pleasant reprieve from the intensity of their earlier discussions. The countryside of Oxfordshire unfolded like a painting, with rolling green hills, quaint stone cottages, and winding roads flanked by ancient oaks, offering a soothing backdrop for their drive. Inside the car, Jack and Bella held hands the whole way there, and their conversation with Charles was light and filled with light-hearted small talk. They chatted about the scenic beauty of Oxfordshire, the upcoming events at Bentley Hall, and Uncle Andrew. Charles, ever the attentive listener and speaker, filled the air with updates about the estate, local news, and occasional anecdotes from his recent drives. The familiarity and ease among them made the journey not just a physical transition from one place to another but also a mental shift from professional concerns to personal comfort and the safety of being amongst trusted company.

As the car turned a final corner, Bentley Hall emerged, a stately masterpiece nestled amidst manicured gardens. Jack had heard about the estate but never expected the well-maintained splendor not only of today but also of elegant times past. The car swept through the gates and up the long driveway, flanked by manicured gardens. Somehow, Andrew and Bea, alerted to their arrival, were waiting by the front entrance with beaming faces. For Jack, the grandeur of Bentley Hall was rivaled only by the warmth of the welcome he received. He couldn't help but feel both humbled and elated to be embraced so openly into Bella's world. Bea embraced Bella tightly while Andrew greeted Jack with a firm handshake.

"So wonderful to meet you," he said.

"Bella has gushed about you," Bea teased, winking conspiratorially. Bella rolled her eyes and nudged Bea gently.

"Stop it, Bea," she protested, though her laughter betrayed her amusement.

Amidst the greetings, Henry, the beloved family pug, couldn't contain his excitement. He scampered up to Bella, his entire backend wagging vigorously, and she scooped him up, showering him with affection and nuzzling his wrinkled nose.

"Oh Henry," Bella said cheerfully. "You always know how to make a girl feel welcome."

Andrew formally welcomed Jack to Bentley Hall and invited them into the library's seating area, where plush leather chairs and a comfortable couch awaited them. The library was a sanctuary of warmth, with towering shelves of leather-bound books, the crackling fire casting a soft amber glow across the room, and the scent of aged paper mingling with the smoky aroma of fine Glenmorangie 18 Year Scotch whiskey. Drinks and snacks were thoughtfully arranged on a central coffee table. Crystal glasses and a decanter of the Glenmorangie whiskey, promised a leisurely welcome.

Settling into the deep cushions, they continued their conversations and gently bathed in the fire's glow. This perfect blend of casual elegance and heartfelt hospitality made it clear that Jack was not just a guest, a cherished friend welcomed into a home filled with love and history. Bella and Jack shared the story of their reunion in Geneva, and Bea shared the young Bella's experience on the Camino and affirmed the validity of her 'love at first sight'.

After the informal dinner in the library, Bella excused Andrew and Bea, using the excuse of Jack being tired from a day of intercontinental travel, which was only partially true.

She led Jack up the grand staircase of Bentley Hall to show him to the blue guest room, reserved for visiting dignitaries. It was elegantly appointed with rich, royal blue draperies and matching accents that complemented the soft, cream-colored walls. It was a large room with a sitting area, a small private library and of course the welcoming bed. The furniture was of classic design, with a large, comfortable bed taking center stage.

As they entered the room, Bella, with a playful and affectionate glance towards Jack, kissed him and whispered, "I can't wait to get you back into bed." Her words, filled with anticipation, sparked a shared excitement, deepening the connection between them.

Jack decided to freshen up, heading straight to the en suite for a refreshing shower. The suite was stocked with plush towels and luxurious bath products that added an extra touch of comfort to their stay. Jack, seizing the romantic opportunity, invited Bella into the shower. In the privacy of the en suite, Jack and Bella found a tender moment to reconnect, the warm cascade of water washing away the fatigue of their journey and deepening the bond they had missed during their time apart. Steam filled the room and they shared intimate moments, rekindling their physical and emotional bond in the warm embrace of the shower's spray. After a brief towel dry, they made their way to the bed and found comfort in each other.

This intimate beginning to their stay at Bentley Hall set the tone for a deeply personal and memorable visit, allowing them both to explore their feelings in the privacy and luxury of the historic estate. Their time together in the blue guest room, away from the world, promised to be not just a physical union but a profound celebration of their reconnection and the life they envisioned together moving forward. As the quiet of the evening enveloped them, Jack and Bella found themselves not just in the comfort of Bentley Hall, but also in the realization that their future, once uncertain, now felt brighter and full of promise.

CHAPTER 45
EXPLORING THE ESTATE

Bentley Hall Oxfordshire, England
03 November 09:38 Local Time

Waking to the gentle sound of rain tapping against the windows, Jack and Bella started their day enveloped in the cozy bed quilt. Spooning they felt one another's warmth and excitement. They shared passionate kisses in the warm bed which of course led to Jack being aroused. Before Bella could do anything about it Jack slipped under the covers to find Bella's tight belly which he kissed leading him to her thighs and the welcoming aroma of her womanhood, his fingers found her swollen lips and slid effortlessly into her. He knew her well in the short time they'd been physical lovers and he knew just where to place his fingers and tongue as he raised her excitement. Her rocking hips quickened, and she held Jack in place as she climaxed. She then let him go but only long enough to pull him up and slide him into her for a mutual orgasmic ecstasy.

"Jack, I'll let you wake me up every day!"

Jack answered, "I would give anything to be able to do so." He relished not only the moment but the fact that they had serendipitously reconnected. He was lost in his own bliss, her scent and her comfort.

Bella brought Jack back when she replied, "Be careful what you wish for. If I get to have you, I'll never let you go." She believed that she and Jack had the same hope for the future.

Jack smiled, his gaze softening. "The stars brought us together for a reason, and I think they're still guiding us."

They shared a shower, followed by the usual morning ritual of brushing teeth and hair, then descended to the morning room for breakfast amid the ambiance of Bentley Hall. Arrayed across the antique side table in the dining room was a traditional English breakfast—eggs, bacon, sausages, tomatoes, mushrooms, and black pudding, accompanied by toast and tea. The rain-kissed Oxfordshire landscape, under gunmetal gray skies, only deepened Bentley Hall's historic charm.

"We're going to Oxford today for some shopping with a definite stop at Blackwell's Books," Bea said. "Would the two of you care to join us?"

Bella and Jack looked to one another, then almost in unison, "No thanks." They both laughed before Bella continued, "We haven't had much time to understand Jack's research together. This is a business trip! We do have to make the most of our time today. I think from the preliminary review, it's very exciting. Groundbreaking."

"Well, in that case, we'll take the two-seater and make it a sports car day. If the sun comes out, we may be able to remove the top." Andrew said. A very personal smile on his face as if he'd been looking forward to this for a while.

"Enjoy, we'll let you know if we will be back for dinner," Bea added.

After breakfast, they retreated to the library, a room filled with natural light despite the overcast sky, where they could spread out Jack's research materials. Bella, with her keen scientific mind and expertise, delved into Jack's findings. Together, they poured over the data and experimental results that Jack had meticulously compiled.

As Bella reviewed the information, she confirmed Jack's initial observations—the new crystalline structure he had synthesized was indeed emitting extra neutrons, a phenomenon both puzzling and exciting, a reaction that produced pure energy at room temperature. This discovery hinted at a potential breakthrough in energy production, the implications of which were profound, as it suggested a viable path towards achieving fusion energy, possibly even cold fusion, which had long been considered the 'holy grail' of energy science due to its promise of providing a clean, almost limitless source of power.

The realization of what this could mean for humanity was staggering. If validated and harnessed, this new form of energy could revolutionize not just the technology of energy but reshape global economies, environmental strategies, and even geopolitical dynamics. It had the potential to change every priority of mankind, from how to power cities to how to approach climate change.

Excited and energized by the scope of their discovery, Jack and Bella began to discuss the next steps. They needed to conduct further experiments to replicate the results consistently and to begin theorizing about the mechanisms at play in the new crystalline structure. Jack placed a small metal box, the size of a ring box, on the table - the box containing the crystalline sample.

Bella's eyes lit up, a playful grin spreading across her face. 'YES, YES, YES, I'll marry you!' she exclaimed, her laughter filling the room.

Jack was paralyzed until Bella burst into laughter.

"If I had known it would be a 'yes', I'd have brought a ring too," Jack said, finally laughing himself. "Now that I have the answer, let's go into Oxford and pick out the ring."

"Jack as much as I want to be married to you, as scientists, we have work to do. We can arrange an appointment with the family jeweler, each pick out rings, maybe a pair of matching earrings too. For me, not you!"

Jack pinched his tongue between his teeth briefly. "I had a feeling this was going to be a life-changing trip. So little did I realize."

As the rain continued to fall outside, their work continued in the atmosphere of the realization that their lives were changing. Bella realized that Jack's work was on the brink of something truly transformative. One would see something interesting, then the other.

"Jack, it surely looks like you've done it. If this bares out it will be the finding of the century, even the millennium. We can go to my lab using the sample you brought and replicate your findings. Shall we go today?" Bella asks.

"There's more I want you to see in these results. Let's finish our work then, given the time, we can either go today or tomorrow. We can also see if the jeweler is available. I don't think anything will change with either result."

"Okay, I'll text the jeweler. I'll have a lab assistant set up replication of the proof."

Their engrossing discussion about the future was abruptly shattered by a clatter from the entry foyer, pulling them back to reality. Charles's voice, raised in argument, was quickly followed by a worrying thud that resonated through the grand hallways of the estate. Bella was about to run to the grand hall to see what help Charles might need when Jack grabbed her arm, "Wait, be careful!" he said.

Muffled voices grew closer, accompanied by the thud of boots on the polished floor. Jack's pulse quickened as the chilling words finally reached their ears: "Check everywhere. We don't leave without them."

Bella, with a mix of fear and quick thinking, peeked through the library door. She saw Charles seated groggily on the floor with his hands zip tied. The intruders, black-clad and wearing balaclavas, moved into other rooms. She ducked back.

Realizing the immediate danger, she told Jack to gather all their research—documents that could not fall into the wrong hands under any circumstances. The sample was already in Jack's pocket. As Jack hurriedly complied. Bella grabbed Jack's hand, pulling him toward a section of the library wall. Her fingers trembled slightly as she pressed the concealed latch, the faint click of the mechanism echoing in the tense silence.

Years earlier, Bella and Bea had discovered a secret door in the wall of the library—a remnant of the estate's long and storied past, perhaps originally intended as a servant's passage or a hidden retreat during tumultuous times. With swift movements, familiar from previous playful explorations, Bella pressed a concealed latch, and a section of the book-lined wall swung open silently.

They quickly slipped into the narrow, dark passage, and pulled the secret door closed behind them just as an intruder entered the library. The intruder paused, taking in the abandoned scene. His voice carried a note of certainty as he called out to his accomplices, "They were here, they have to be nearby!"

Inside the passage, Jack and Bella listened to the conversation occurring on the other side of the panel. The intruders were looking for them and them only. As silently as they could, they retreated further away from the scholarly atmosphere of the library. The realization that their groundbreaking discovery had been compromised hit them like a tidal wave, the weight of its implications pressing down on them as they navigated the dim passage. The passage wound its way through the inner walls of the estate and led them to the steps down to the cellars.

Inside the dark, narrow passage, Jack and Bella moved carefully, the air thick with dust and the faint smell of old wood. Their footsteps were barely audible against the stone floor as they descended toward the cellar, Jack clutching the precious crystalline sample in his pocket. Bella's steady grip on his arm reminded him they were in this together, no matter how perilous it seemed.

They needed to think beyond just escaping. They needed to secure their research and possibly contact the authorities or their own security for help. The implications of their discovery—and the apparent knowledge of it by nefarious parties—added weight to every decision. Their world had just become a hell of a lot more dangerous.

CHAPTER 46
INTRUDERS

Bentley Hall Oxfordshire, England
03 November 14:02 Local Time

The intruders, equipped with advanced gear, worked with precision.

One of them brushed his hand against the still-warm coffee mugs, his voice low but firm as he reported, "They're close." Spurred by this discovery, the team intensified their search. Two members methodically tapped the library walls, hunting for hidden exits—a standard procedure in such historic estates. The others spread out into adjacent rooms, their movements deliberate and silent, leaving no corner unchecked.

In the grand hall, Charles remained under watch, his hands zip-tied and his composure unwavering. He resisted the urge to glance toward the library, aware that any indication of the secret passage could endanger Jack and Bella. His only weapon was silence.

Inside the hidden passage, Jack and Bella moved cautiously. The air was thick with dust, the confined space amplifying the sound of their shallow breaths. Bella, familiar with the passage from childhood explorations, led the way toward the cellars. Her mind raced, piecing together the intruders' intentions and calculating their next move.

Jack followed closely, his grip tightening around the crystalline sample in his pocket. Despite the darkness, his Navy training kicked in. He scanned every corner, listening for the faintest noise that might betray their pursuers. "How far to the cellar?" he whispered.

"Not far," Bella replied. "We'll head for the weapons cache first."

The passage led them into the cool, dimly lit cellar. Knowing the layout of Bentley Hall intimately, she was aware of several options. The passage to the armory, with its historical and valuable antique weapons displayed prominently, along with modern high tech hunting weapons was one potential refuge. The armory also held a secure gun safe with high caliber sniper rifles equipped with thermal sights, semi-automatic pistols with tactical modifications even a semi-automatic shotgun all for the defense of the estate. It was loaded with modern weapons but required time and access codes. Instead, Bella chose the cellars as their destination. She remembered the additional cache of hunting weapons and handguns Andrew had discreetly hidden there—a precautionary measure from years past, now a potential lifeline.

Bella quickly located the concealed cupboard and retrieved two handguns. She handed one to Jack, who checked it with practiced ease. "We haven't talked about my Navy days," he murmured. "But I know my way around these."

Bella's brow arched, but she wasted no time questioning him. Instead, she grabbed a hunting rifle with a high powered scope and two tactical knives. "Five of them," Jack noted quietly. "One guarding Charles, four searching for us. They'll be down here soon." "We'll use the tunnels," Bella said, her voice steady despite the urgency. "They connect to the estate grounds. If we time it right, we can double back and ambush them on their way out."

Jack nodded, impressed by her calm decisiveness. "Let's go."

As they moved deeper into the cellar, faint creaks and distant footsteps echoed above. Bella led them to a dust-covered cabinet, pushing it aside to reveal a small, hidden door. Beyond it lay the old wine tunnels, their walls lined with cobwebs and faintly glowing from patches of moss.

"Through here," Bella whispered. She glanced back at Jack, her confidence unwavering. "We'll be fine."

Jack met her gaze, her determination bolstering his own. "I trust you," he said simply.

They stepped into the tunnel, the cool air brushing against their faces like a promise of safety. The faint hum of a car engine rumbled through the walls—a sound that could herald rescue or another wave of danger. For now, they pressed on, hand in hand, their resolve unshaken.

As they disappeared into the shadows, Jack's fingers brushed the crystalline sample in his pocket. Whatever awaited them on the other side, he knew their discovery—and their bond—was worth fighting for.

CHAPTER 47
THE BATTLE AT BENTLEY HALL

Bentley Hall Oxfordshire, England
03 November 14:17 Local Time

Jack and Bella, armed and resolute, moved cautiously through the damp, dimly lit cellars of Bentley Hall. Their mission was clear: secure an escape, contact the authorities, and protect Charles, who remained vulnerable upstairs. Every sound echoed ominously in the confined space, each creak of the ancient stone walls a reminder of the imminent danger.

Then Bella heard the voices within the tunnel - the intruders were closing in. Their voices, methodical and sharp, carried through the space, accompanied by the static squelch of radios. Bella's sharp ears caught the sound, her pulse quickening. They were sweeping the rooms, coming closer.

"There's a recess behind those barrels," Bella whispered, pointing to a shadowed alcove. It wasn't ideal, but it could buy them time.

Jack, however, had other plans. His jaw set, he glanced at the doorway they had just passed through. "No," he murmured. "We take them here."

Bella hesitated but trusted his judgment. "If we keep it quiet, we might manage this," she said, gripping her knife tightly.

Jack nodded, grabbed his blade for himself and tested its edge. Razor-sharp. They moved into position, Jack bracing behind the door while Bella melted into the shadows, her training and instincts guiding her. Jack acted as if he'd done this before.

Each moved into their position of stealth. The voices grew louder, the beam of a flashlight slicing through the darkness. Jack steadied his breathing. Bella's grip tensed on her blade.

The first intruder stepped into the room, sweeping his weapon and flashlight across the walls. Bella held her breath, unmoving. As he passed her hiding spot, she sprang into action. With one precise movement, she sliced through his carotid artery. Blood sprayed in a violent arc, drenching her face and hands. The man collapsed, gurgling his last breath.

The second intruder froze in the doorway, staring in horror at his fallen comrade and the blood-covered woman standing over him. His weapon swung toward her, but Jack struck before he could fire. With the speed of a predator, Jack twisted the rifle upward, grappling with the man as suppressed shots ricocheted harmlessly off the ceiling. Holding him steady, Jack gave Bella the opening she needed. Her blade flashed again, cutting through the second man's throat. He fell, lifeless, joining his partner on the floor.

Bella wiped the blood from her face, her hands trembling. "They're... gone," she whispered.

Jack nodded grimly. "Search them. Radios, weapons—anything useful."

They worked quickly, turning down the radios' volume and pocketing them. The intruders carried no identifying markers—no clues as to who had sent them.

Jack's voice was tight. "These knives are so sharp. Did Charles and Andrew anticipate this?"

Bella allowed himself a faint smirk. "Paranoia or foresight—it doesn't matter now. Let's move. Two down, three to go."

As they slipped through the cellar's twisting tunnels, Jack and Bella moved cautiously, their every step muffled on the stone floor. The radios crackled intermittently, fragments of conversation breaking through.

A commanding voice barked through the radios, laced with frustration, "I want them found—now! We don't have more time."

"They're on a schedule," Bella murmured, her jaw tightening. "They're here for us."

"And Charles," Jack added. "If they leave, he's the only one who can identify them."

Bella swore under her breath. "Then we can't let them leave."

"Listen," Jack said, gripping her arm gently but firmly. "We've got this. Quick and quiet."

Bella nodded, her breathing steadying. They reached the cellar's exit and carefully emerged into the estate grounds. The day was cold, and wet, the air sharp and filled with tension. From their concealed position behind the garden shrubbery, they observed the remaining intruders. The individual guarding Charles stepped outside, clearly agitated, and communicated over the radio. His urgency transmitted across the open space, his voice was sharp and commanding as he ordered a retreat. The response from the others affirmed that they were preparing to leave, mentioning a lack of communication with the team in the basement and their intention to take Charles with them. Bella closed her eyes, hoping to focus her hearing on the instructions. Fortunately, or rather, not, she heard mention of Charles, as he barked into his radio.

"What about this old man?" came a voice from the radio.

The guard hesitated. "Kill him. He'll slow us down."

Bella's breath hitched, misting in the cold air. "We're out of time," she whispered.

Jack steadied the rifle he had taken, his hands calm and practiced. "This scope better be right. I've got this. Be ready."

Bella watched as he exhaled slowly, the barrel of the rifle aligning with the guard's head. The silence created a question of whether Jack would take the shot. The trigger clicked, followed by the crack of the bullet through the air. It caught the guard in the center of his head and brought him down instantly. The guard crumpled, lifeless, to the ground.

The remaining intruders, startled by the gunshot, bolted for cover. Jack and Bella stayed low, listening as the intruders argued in hushed tones. "They've got a sniper!" one hissed.

"Screw this. Get to the car!" shouted the second.

Jack and Bella remained hidden as the men dashed toward a waiting vehicle. The engine roared to life, gravel spraying as they sped away into the night.

Wasting no time, Jack and Bella rushed into the grand hall. Charles sat bound and pale but alive.

"Hold on," Jack said, cutting through the restraints. Bella grabbed her phone, dialing 999 - emergency services. Her eyes scanned the doorway, vigilant for any lingering threats. Within seconds, Charles was free. After the 999 call, Bella called Andrew, relaying the events with urgency.

Immediately following their call, Andrew called his contacts within the Home Secretary's office and had armed private security arranged to arrive at the house as soon as possible.

By the time emergency services arrived, armed and ready, Jack and Bella had fortified the estate as best they could.

Bella retrieved the crystalline sample from Jack's pocket, her expression one of quiet determination. "We protect this. No matter what." Jack nodded. "And we find out who's behind this. Together."

As the sun set over Bentley Hall, casting a pale light over the battered estate, a sense of calm began to settle. They had survived the ordeal, but the fight was far from over.

Jack glanced at Bella, his voice steady. "We make a hell of a team."

She allowed herself a small smile. "Yes, we do. And we're not done yet."

Hand in hand, they stood together, ready to face whatever came next.

CHAPTER 48
THE AUTHORITIES

Bentley Hall Oxfordshire, England
03 November 16:17 Local Time

Officials swarmed the property. Armored vehicles ambulances, and police cars with blue lights flashing cut through the tranquility of the Oxfordshire countryside. Bella, composed despite the recent events, told Jack to let her take the lead in communicating with the authorities. She quickly identified herself, Jack, and Charles, and three other staff members who had been on the estate who were unharmed during the ordeal. She clarified that they were all safe and the blood they were covered with came from the bad guys. They agreed to gather at the front entry to assist in the authorities' efforts to secure the estate.

Specialist armed teams methodically moved through Bentley Hall, clearing each room. While other teams cleared each section of the rolling and wooded grounds of the property. They confirmed no additional threats remained and took care of securing the scene and ensuring that all evidence was tagged and bagged for the forthcoming investigation.

After the initial crisis management and securing of Bentley Hall by the MI5 and local task force, the focus shifted to detailed evidence gathering and taking individual statements from Bella, Jack, and Charles, which they did in separate rooms. Bella, recalling the events, felt adrenaline still surging through her, the memory of just moments ago keeping her nerves on edge. Part of her was uncomfortable letting the authorities seemingly ransack her home for evidence.

The investigators collected physical evidence from Bella, Jack, and Charles to rule out any inconsistencies—securing their clothing and taking DNA and fingerprint samples.

The investigators ensured that this process did not compromise any potential evidence. They then were allowed to shower, clean up, and change into clean clothing, which helped them begin to recover physically and mentally from the ordeal.

Once they were composed and dressed, Bella, Jack, and Charles sat together in the lounge—the library was no longer the place of comfort and calm they wanted—they talked and drank together quietly whilst listening to the footsteps of officers working the scene. The only movement from any of them was for Jack to reach for Bella's hand.

Andrew and Bea returned to Bentley Hall, calm but frustrated by the security checkpoints Once cleared, they were allowed to enter their home, where Bella awaited them to provide them with the details. They were shocked as they learned of the events of the manor in their absence.

With the statements taken, the bodies removed, and the bulk of the emergency vehicles taking their leave, the estate slowly returned to a semblance of normalcy following the intrusion. Initially the arrival of law enforcement and emergency services marked the transition from immediate danger to recovery. Now with the bulk of the emergency vehicles taking their leave the real work of understanding who and why begins.

Taking charge, Andrew's priority was to ensure the safety and well-being of his family and staff. He organized a quick debriefing to account for everyone's physical and emotional state and to understand fully what had happened in his absence. His leadership helped reassure the shaken staff, restoring a sense of order and safety to the household. He exudes British command.

As the evening drew on, the estate began to feel like a home again, albeit one touched by recent shadows. Armed specialists remained to provide a level of security. Discussions about future security measures and the implications of the attack were postponed for later meetings.

Andrew informed them that a team, a task force comprising MI5 and MI6 agents, was on the way via helicopter, seeking a direct and secure meeting with Bella, Jack, and Charles. Bella and Jack held back their surprise, but the idea that their work had warranted both the attention of intruders and the security of such scale bewildered them. Sure, Jack's research could well lead to staggering technological advancements, but somehow had it reached the ears of antagonist groups?

Andrew, well-versed in the protocols of handling such high-level security matters, agreed to facilitate the task force's helicopter arrival, an indication of the urgency and importance of the meeting and offered the set-up of a temporary command center on the estate grounds. He also, extended his hospitality by offering accommodations to the security and task force personnel for the duration of their need. This gesture not only provided logistical support to the task force but also reinforced a cooperative relationship between the estate and the security services, ensuring that all actions could be carried out with minimal disruption to ongoing estate operations and the daily lives of its residents.

The atmosphere at Bentley Hall shifted from one of recovery to one of efficiency and a beehive of activity; the household knew how to prepare for guests. Everyone busily prepared for the arrival of additional occupants as if preparing for a royal visit. This certainly gave the staff a focus on work, not on worry.

When they arrived, the MI5 leadership assured Andrew that they would take over the responsibilities of armed security. MI5 and MI6 team members took charge as if the king himself were coming. Bella and Jack, in a way, did feel enveloped in a measure of the security that was close to overwhelming, but they definitely couldn't say that they weren't now objectively safe.

This collaboration highlighted the critical nature of Jack and Bella's work and the lengths to which various conspirators will go to disrupt or seize the innovative scientific breakthroughs. The upcoming presence of such high-level agencies underscored the national security interests at stake and set the stage for a comprehensive plan to address the forces behind the intrusion.

CHAPTER 49
NATIONAL SECURITY

Bentley Hall Oxfordshire, England
04 November 07:17 Local Time

After a night of guarded rest, Bella and Jack found solace in each other's company amidst the ongoing tension at Bentley Hall. The estate had been temporarily transformed into a high-security compound, bustling with activity. MI5 and MI6 personnel worked methodically, their presence creating an atmosphere reminiscent of wartime preparations. Such a level of government involvement hadn't been seen in the Cotswolds since World War II—a sobering reminder of the gravity of the situation.

The morning began with a structured routine. After a quick breakfast, a critical debriefing was held with key investigators, Andrew, Bea, Bella, and Jack. The atmosphere was solemn, the air heavy with urgency. Everyone present was sworn to confidentiality under the Official Secrets Act, a formality that underscored the sensitivity and importance of the information they were about to receive.

During the briefing, it was revealed that the intruders were not common criminals but Iranian commandos from the elite

Quds Force. This shocking disclosure elevated the incident to an international crisis, linking it to state-sponsored actions and raising the stakes considerably. MI5 shared intelligence on an Oxford-based cell, part of a broader network, that had been under surveillance. The intrusion was evidently a coordinated effort. While some members of the cell had been apprehended in a swift operation, it was unclear whether the two escaped intruders were among those captured.

Investigators probed deeply, attempting to uncover the reasons behind the attack. The group examined every angle, from personal histories to professional connections, seeking any clue that might explain why Bentley Hall and its occupants were targeted. Jack, aware of the potential global ramifications of his research, suggested involving the FBI. He was reassured that the FBI had already been contacted and that two task force members familiar with the break-in at his lab were en route.

Jack took the opportunity to share details of his recent research, speculating that its groundbreaking nature in quantum computing and material science had likely drawn the attention of the operatives. His breakthrough—a potential game-changer in energy production—could be a prime target for foreign powers seeking technological dominance. While the agents were familiar with the general implications of quantum research, they were not fully prepared for the depth of Jack and Bella's findings. The scientists conducted an impromptu tutorial, walking the group through the fundamentals of quantum mechanics. Bella, drawing on her experiences at Oxford, explained complex concepts with clarity and patience. The session extended well beyond its scheduled time, stretching through lunch and into afternoon tea. By the end, the team had a solid grasp of the stakes and a renewed sense of the weight of their responsibility.

Meanwhile, Bentley Hall buzzed with a mix of tension and efficiency. Under Charles's adept management, the estate adapted seamlessly to its temporary role as a command center. Household operations continued with precision. As dinner approached, the staff prepared a traditional English Sunday supper: roast beef, Yorkshire pudding, roast potatoes, and fresh vegetables. The hearty meal provided a brief but welcome reprieve from the intensity of the day, allowing everyone to regroup and recharge.

The arrival of two American FBI agents, Sergeant Berriman, who had been reassigned to the FBI and Special Agent in Charge Mickey Golden, coincided perfectly with dinner. Their presence underscored the international cooperation at play. While the agents politely declined alcoholic beverages, sticking to non-alcoholic options during the meal, Bella, Jack, and the household allowed themselves the modest relaxation of wine, balancing the social aspect of the meal with the gravity of their situation.

Once dinner concluded, the focus returned to the pressing matters at hand. Sgt. Berriman provided an update on the situation at the University of Arizona, detailing heightened security measures implemented at Jack's lab. The revelation that a postgraduate assistant had disappeared added a chilling layer of complexity. Investigators were working to determine whether she had been compromised or had willingly leaked sensitive information. Her disappearance became a central concern, with efforts intensifying to locate her and understand her role in the breach.

Agent Golden revealed additional measures, including the deployment of the FBI's elite Hostage Rescue Team (HRT) to secure both Jack's university lab and his private mountain laboratory in Arizona. This heightened level of security underscored the gravity of the situation and the government's commitment to safeguarding every aspect of the research.

Discussions stretched late into the night, with every agency sharing insights and coordinating strategies. The meeting culminated in a robust plan: immediate security measures, protocols for safeguarding the research, and long-term strategies to counter future threats. Tasks were assigned, and timelines established, the collective resolve of the group evident in their unwavering focus.

As the evening deepened, the gravity of the situation became ever clearer. Bella and Jack exchanged a glance, their connection strengthened by the shared weight of their discovery and the danger it had brought. They had faced the immediate threat together—and now, they would face the battles to come with the same unity and determination.

CHAPTER 50
THE DECISION

Bentley Hall Oxfordshire, England
05 November 08:02 Local Time

As they sat at the breakfast table, the warm morning light spilling through the windows, Bella and Jack found a brief moment of tranquility amidst the chaos of recent days. Jack's thoughtful gaze lingered on the light before he broke the silence.

"You know," he began softly, "last night and even now… these moments with you are what keep me grounded through all this madness."

Bella reached across the table, taking his hand. "It's the same for me, Jack. Everything that's happened—it's terrifying, but having you here makes it all manageable."

Jack's face grew more serious. "Do you think I've pulled us into a life where we'll always have to look over our shoulders?"

Bella laughed lightly, her grip on his hand steady. "I don't think so, Jack. But I have been thinking about how we can protect ourselves and make sure we're safe without living in constant fear."

"Great minds think alike," Jack replied, his lips curving into a faint smile. "I was wondering... what if we moved our base to my place in Tucson? The lab there is underground, secure in an old mine shaft. It would be safer for us to work without endangering anyone else. But I know how important your family and Oxford are to you, and I don't want to take that from you."

Bella drummed her fingers thoughtfully on the table. "The labs at Oxford are incredible, but... Tucson might give us the security we need. She stared for a moment, trying to produce a picture of Tucson. Then added "Henry would have more space to run around than at Magpie Lane and I think he's always liked the mountains."

Jack chuckled, watching her nuzzle Henry's snout. "Exactly. Plus, with the HRT on-site, we'd have immediate protection. We wouldn't have to worry constantly about safety, and the privacy there is unmatched. We could focus entirely on our work—and the views aren't bad either. From the deck, you can see Kit Peak's telescopes sixty miles away, the Tucson Valley sparkling at night, and even the mountains stretching all the way to Mexico. It's peaceful, and the pool is usable year-round."

Bella's expression softened into something resolute. "We could thrive there, Jack. We'd make breakthroughs without looking over our shoulders every minute. And don't worry—I won't demand top billing on the Nobel Prize."

Jack grinned. "You'd deserve it. Madam Curie has nothing on you."

Bella laughed and leaned back in her chair. "I'm in. Let's do it. If we're going to face threats, I'd rather do it knowing we're on home turf, with the best defenses."

"Together." Jack's voice was warm, and his hug was firm, sealing their decision. "We'll coordinate with the security teams and prepare everything for the move. Do you think Andrew and Bea will understand?"

"They'll understand. Afterall I am a grown woman with my own life. I'll never fully leave Bentley Hall." Bella reassured him. "And we'll keep pushing forward with our work. We're not just going to hide, Jack—we're going to thrive."

Jack pulled out his phone, showing her photos of the estate. Nestled on eight gated acres in the pine-covered slopes of Mt. Lemmon, it blended rustic charm with modern luxury. "It's at 4,500 feet," he explained, "built on an old mine site. The house itself is a mix of historical character and contemporary comfort, with picture windows framing views that change with every season."

Bella's eyes lit up. "It sounds like a sanctuary. And the tunnels—what are they like?"

Jack's grin widened. "They're not just historical. They're practical. Part of my lab is set up there—it's naturally isolated and stable. Perfect for our work. With some funding, we could fully outfit it with whatever equipment we need. It's a clean room lab environment, but underground. What we don't have I'll apply for funding to fully outfit our needs. I think we will have no problem finding partners and sponsors."

Bella's tone turned cautious. "After everything that's happened, I think our biggest partner in this will be the government. They'll love the added security layers. And while we might want to publicize our work, I have a feeling we'll face some censorship."

Jack nodded. "That's fair. But the privacy and focus we'll gain there will be worth it. It'll be more than a lab—it'll be our fortress, our home."

Bella leaned closer, her excitement growing. "Jack, it's perfect—not just as a retreat but as a place where we can focus on our research and live comfortably. It's what we need right now."

They both smiled, the decision solidified. Bella could feel the excitement building, despite the looming threats. This move wasn't just about safety; it was about starting a new chapter together, one where their work and their bond could flourish without constant fear.

With a renewed sense of control and determination, they began planning their next steps. Together, they were ready to face the challenges ahead—stronger, more united, and with a vision of the future they would build hand in hand.

CHAPTER 51
MAKING IT HAPPEN

Bentley Hall Oxfordshire, England

05 November 14:02 Local Time

The library at Bentley Hall, with its warm lamplight and the rich scent of aged leather-bound books, had always been a sanctuary. Tonight, it would also be a place of transition, a space where bonds would be tested and affirmed. Bella and Jack, seated close together on the old, tufted two-seater couch, felt the significance of the moment. Andrew and Bea, their ever-reliable confidants, occupied their usual club chairs. There was a solemnity in the air, though it was softened by the familiarity of the room and the unspoken trust they all shared.

Jack poured Andrew's favorite Scotch—a rare Highland Glenmorangie 18-year-old, with its light but refined style and hints of honey and figs—something Jack had come to appreciate during their evenings of conversation. Perhaps it was the fact that it's aged in both bourbon and sherry casks. Jack, of course, still preferred his favorite bourbons. He handed glasses around, starting with Bea and Bella, before finally taking his seat next to Bella. The clinking of crystal was a silent acknowledgment of the changes to come, and the first sip settled the weight of the announcement they were about to make.

Jack took a deep breath, his voice calm but steady. "Andrew, Bea, we wanted to sit down with you both tonight because we have something important to discuss. It's about our future, both personally and professionally."

Andrew leaned forward slightly, resting his elbows on the arms of his chair. "This sounds serious," he said, his tone measured but curious.

Bella placed her hand over Bea's, a small gesture of reassurance. "It is serious, but we want you to know this decision wasn't made lightly. We've decided to move to Jack's estate in Arizona. It's a safer place for us, given everything that's happened, and it's also better equipped for the kind of work we need to do."

Bea's brows knitted, her lips tightening as she processed the words. "Arizona? Bella… are you certain about this? Bentley Hall has always been your home. You've built so much of your life here."

"It will always be my home," Bella replied earnestly, her voice tinged with emotion. "But after the recent events, Jack and I need a place where we can focus on our work without constant fear. His estate is secure, isolated, and it gives us the privacy we need to continue our research safely."

Andrew sighed, a bittersweet smile softening his face. "I can't say this comes as a surprise. The two of you have been thrust into extraordinary circumstances, and your work—it's world-changing. Bella, you've always been like a daughter to us, and while the thought of you leaving pains me, I understand. You're a remarkable woman, and you've found a man who clearly sees that too."

Bea's composure wavered, her eyes glistening as she spoke. "You know how much we'll miss you, don't you? Both of you. And Henry, of course." Her voice broke slightly at the mention of Bella's beloved pug, who was curled up on a nearby rug, blissfully unaware of the heavy conversation.

"Of course we know," Bella said, her voice steady even as her eyes shimmered. "And this isn't goodbye forever. We'll visit as often as we can. Bentley Hall is part of who I am—it always will be."

Jack nodded, his gratitude evident. "Your support means the world to us. This decision isn't just about safety; it's about giving us the chance to focus on what matters most—our work, and the life we want to build together."

Andrew leaned back in his chair, his expression thoughtful. "The life you're building together—what does that look like in Arizona? What's this estate like?"

Jack smiled, sensing the opportunity to ease the tension with details. "It's tucked away in the mountains outside Tucson, surrounded by pine forests, gated with a perimeter fence, all the modern security electronics. The main house is rustic but modern, with a lab I've set up in the tunnels of an old mine shaft. It's private, secure, and peaceful. We'll be able to work without distractions, and there's plenty of space for Henry to run around."

Bea raised an eyebrow, her curiosity piqued. "Tunnels in a mine shaft? That sounds… unconventional."

Jack chuckled. "It is, but it's also incredibly practical. The underground environment is stable and secure—ideal for sensitive research. And Bella will finally have her own library she's always wanted. We can even expand it together."

Bella's eyes lit up. "And the views are breathtaking. It's not Bentley Hall, but it's beautiful in its own way. I think you'd both love it."

Andrew exchanged a glance with Bea before nodding. "It sounds like the two of you have thought this through. And as much as we'll miss having you here, we want what's best for you. Just promise you'll keep us in the loop—and that you'll come back to visit."

"Absolutely," Bella said, her smile resolute. "This isn't goodbye. It's just a new chapter."

The conversation drifted into lighter topics, memories of shared moments at Bentley Hall, and plans for the move. Though the evening was tinged with sadness, it was also filled with a sense of hope and purpose. As the glasses were refilled and the fire crackled softly in the hearth, the four of them found comfort in knowing that their bond was unshaken, no matter the distance that lay ahead.

The next steps were equally challenging. They had to coordinate with the combined task force of MI5, MI6, and the FBI, who were overseeing the transition to ensure every aspect of the move was secure. Andrew, leveraging his extensive connections, took the lead with the British agencies, his influence ensuring that he stayed informed of the progress and evolving security measures. His steady hand brought a sense of calm to an otherwise daunting process.

In a high-level meeting filled with agents, security experts, and intelligence personnel, Jack and Bella laid out their plans, presenting Jack's Arizona estate as the optimal site for their ongoing research and safety. The estate's isolation and state-of-the-art lab would provide both a sanctuary and a workspace.

"Our home in Arizona," Jack said, and Bella couldn't help but smile at his choice of the word 'our,' "not only offers us the privacy we need but is equipped with a lab that will significantly benefit our research. With the guidance of our advisors, we feel this is the safest course of action."

Agent Golden, a seasoned professional whose hardened demeanor belied her genuine concern, responded with a nod. "We've reviewed the security measures already in place at the Arizona site, and with a few enhancements, we're confident in securing a seamless transition. The FBI's Hostage Rescue Team will provide close protection until we're fully satisfied with the setup. Afterward, we'll transition to an on-site security force. The clearing on the property, currently used by mountain search-and-rescue helicopters, will also allow us quick response options for efficient transportation of equipment, rapid travel, or even emergencies."

Henry Kingston, the commander of MI5 and an old friend of Andrew, added his voice to the conversation. "From a logistics standpoint, MI5 will work closely with your team to secure all sensitive materials in transit. Both our governments are fully committed to supporting this move." His tone carried a weight of reassurance that was as comforting as it was resolute.

The discussions delved into meticulous plans for transporting sensitive research materials and personal belongings while ensuring every step was executed with discretion and top-tier security. Each detail was scrutinized, contingencies outlined to preempt potential threats. Jack and Bella were acutely aware of the lengths everyone was going to, ensuring their safety at every juncture.

As the meeting drew to a close, Jack and Bella shared a quiet moment of gratitude for the professionalism and thoroughness of the task force. Their initial anxieties about the move were tempered by the faith these agents and experts had placed in them—a faith that spanned two nations, reminding them of the global significance of their work.

In the days that followed, as Jack and Bella finalized every detail of their departure, they were filled with a profound sense of purpose and responsibility. The decision to leave Bentley Hall wasn't just a logistical choice; it was a commitment to a new life—a promise to each other, to their research, and to everyone who believed in them. Each meeting, each step of preparation, reinforced their dedication to this path.

Even as they prepared to leave, Bentley Hall remained a wellspring of reassurance. The quiet support of Andrew and Bea was ever-present, a steady reminder of the life they had built there, of the family they would always have. They shared quiet dinners together, lingering over meals, finding comfort in familiar routines. Little Henry's playful barks echoed through the halls, a grounding presence that kept them focused amidst the whirlwind of change.

Jack and Bella's hearts swelled with gratitude as they realized the tireless effort of those working to secure their journey. The international task force—comprised of seasoned professionals from MI5, MI6, and the FBI—had become a shield around them. Each agent and strategist treated the mission not merely as a job but as a cause. It was a collective commitment that deeply moved Jack and Bella.

During quiet moments, Jack and Bella exchanged glances that spoke volumes. They understood the risks and sacrifices but also recognized the incredible network of support surrounding them. It wasn't just about securing a safe place to continue their work; it was about honoring the trust so many had placed in them.

That evening, as they walked the grounds, Bella looked up at the familiar towers and chimneys etched against the twilight sky. "This place… it's given us so much more than a home," she murmured, her voice tinged with emotion. "It's given us family, purpose, and strength."

Jack nodded, his gaze distant yet resolute. "And now it's time to take that strength with us."

As they strolled hand in hand, they felt the bittersweet ache of leaving and the quiet excitement of forging a new chapter in Arizona. This was not just a relocation; it was an evolution—a step into a future defined by courage, resilience, and hope for a discovery that could someday change the world.

CHAPTER 52
RETURN

Bentley Hall Oxfordshire, England
07 November 09:42 Local Time

The visit from the family jeweler marked a poignant moment in their journey, blending personal joy with the gravity of their recent decisions. Bella, radiant with excitement, engaged in meticulous discussions about design, sharing her vision for a ring that symbolized both elegance and their shared bond. After much deliberation, she selected an exquisite engagement ring: a 3.5-carat piece featuring an emerald-cut sapphire flanked by trillion brilliant-cut diamonds, set in platinum.

Jack, typically understated but with a quiet elegance, chose a platinum band with a single square sapphire—a subtle yet striking complement to Bella's choice. As they exchanged glances during the consultation, the symbolism of their decisions resonated deeply. These rings were not merely tokens but tangible representations of the partnership they were building—a union of love, intellect, and purpose.

Their engagement was sealed that evening in a warm family gathering. Charles, always attuned to moments of significance, orchestrated an over-the-top celebratory dinner alongside the kitchen staff. The dining table was adorned with fresh flowers from the estate gardens, and each dish reflected the thoughtfulness and care of a family celebrating its newest chapter. The clinking of glasses, heartfelt toasts, and laughter echoed through Bentley Hall, blending tradition with the promise of a bright future.

Bea, ever the matron of the family, was entrusted with the future wedding planning, a role she accepted with genuine enthusiasm. "The wedding must be at Bentley Hall," she declared, her eyes sparkling with determination. "I'll ensure it's a day worthy of both of you—timeless, elegant, and unforgettable."

Bella leaned over and kissed Bea on the cheek. "I wouldn't trust anyone else with this, Bea. You've made Bentley Hall a home for me, and having our wedding here would mean everything."

Jack nodded in agreement. "We can't think of a better place, but let's focus on the work ahead. We've got some hurdles to clear before the big day."

Bea waved her hand dismissively. "Leave the wedding to me. You two focus on changing the world."

The days that followed were punctuated with moments of calm amidst the whirlwind of preparation. Strolling through the estate grounds with Henry bounding ahead, Jack and Bella found clarity in these peaceful interludes. "It's funny," Bella mused, watching Henry sniff out his favorite spots. "The answers to our biggest challenges always seem to come when we're not actively looking for them."

Jack chuckled, nodding in agreement. "That's the beauty of it. Sometimes, stepping away from the problem is the best way to solve it."

These walks became their sanctuary, a time to reconnect and let ideas flow freely. Amidst the November fall beauty, the changing of the colors, the warm glow of serenity of Bentley Hall, their minds were at their sharpest.

The confirmation of their breakthrough—the generation of extra neutrons from the crystalline material—brought a renewed sense of urgency and purpose. This discovery, with its profound implications for energy, computing, and quantum physics, naturally steered their focus towards the elusive goal of cold fusion. Discussions with leadership at the UK Atomic Energy Authority and their counterparts at the US Atomic Energy Agency solidified their path forward.

"We're entering uncharted territory," Jack said during a late-night brainstorming session. "But this is what we've been working toward. Cold fusion isn't just a dream anymore—it's within reach."

Bella, her eyes alight with determination, added, "And we'll do it safely, responsibly. Zeliconite's properties give us an answer, and the lab in Tucson is the perfect place to push these boundaries."

An agreement was reached and signed between the governments and agencies to collaboratively design and build the necessary equipment at Jack's underground lab. The initiative would remain classified, leveraging the safety and isolation of the remote mountain site. "This isn't just about scientific progress," Bella noted during a planning call with their government partners.

"It's about proving to the world that this can be done ethically and securely."

The logistical aspects of setting up the lab were handled with remarkable efficiency. Government organizations coordinated the transfer of high-tech equipment from Sandia Labs in New Mexico, while Jack's personal contacts ensured the process remained discreet. Helicopters and trucks carried state-of-the-art machinery, computers, and software to the secluded estate, blending seamlessly into the quiet rhythms of the mountain community. Locals, accustomed to Jack's ongoing renovations and improvements, took little notice of the activity.

The pair joined a video call with the team on site in Arizona. "It's almost too perfect," Jack remarked as they watched via the video call, a helicopter lower a crate of sensitive equipment onto the property. "The location, the timing—it all feels like it was meant to be. Thanks folks for all your hard work back home. "

Bella smiled, her hand resting lightly on his arm. "It is perfect, Jack. We're building something extraordinary here—not just for ourselves but for everyone who'll benefit from this work. This is bigger than us, and we're ready for it."

Their confidence grew with each passing day. As the lab began to take shape, it became a symbol of their commitment—not just to each other but to the pursuit of knowledge and the betterment of humanity. In the quiet moments, amidst the hum of progress, they felt the weight of their journey but also the strength of their resolve. Together, they were unstoppable.

CHAPTER 53
TRAVEL

Bentley Hall Oxfordshire, England
09 November 11:33 Local Time

Their time at Bentley Hall had come to an end. The distinct thrum of helicopter blades echoed through the estate as the aircraft lifted off from the large carriage sweep, a historic circular driveway designed to gracefully welcome guests. Now, it was repurposed as a makeshift helipad, signaling the gravity of their departure. Inside the helicopter, Jack, Bella, and Henry sat quietly, the weight of their goodbyes still fresh. Bella's face was streaked with tears as she waved back at Bea, Andrew, and Charles, who stood at the edge of the sweep, their silhouettes etched against the gray English sky.

The helicopter ride to the Royal Air Force base was brief, but its significance loomed large. From there, they boarded a U.S. government private jet, a sleek and secure aircraft staffed with representatives from both nations' atomic and security agencies. The flight itself was anything but ordinary. As the jet climbed to its cruising altitude, the faint hum of fighter jets—RAF Typhoons on one side and USAF F-35s on the other—reminded everyone onboard of the unprecedented security measures in place.

Henry, ever the calming presence, lay curled up on Bella's lap, his head resting on her leg. She absentmindedly stroked his fur, her eyes scanning the cabin. Jack leaned closer, his voice soft. "You okay?" he asked.

Bella nodded, offering a small smile. "Just… processing. Leaving Bentley Hall felt like leaving a piece of myself behind."

Jack reached for her hand. "It's not a goodbye. It's just a new chapter. We'll carry the spirit of Bentley Hall with us—its strength, its memories."

Bella squeezed his hand, finding comfort in his words.

The flight was a mixture of intense focus and moments of levity. Jack and Bella, alongside the governmental representatives, dove into discussions about the cold fusion proof of concept. Detailed schematics, logistical plans, and security protocols were reviewed meticulously. Each person onboard carried a piece of the puzzle, their collective expertise converging toward the groundbreaking project.

Despite the serious nature of their work, the atmosphere was tempered with small comforts. Gourmet meals, carefully paired wines, and seemingly endless cups of coffee kept the team energized. "It feels ironic," Jack mused over a perfectly brewed espresso, "working on something so revolutionary while indulging in such luxury."

Bella smiled faintly. "Maybe it's a reminder of what we're fighting for—a future where everyone can live without scarcity, where comfort isn't a privilege but a norm."

The agents informed them that the Arizona lab was progressing rapidly toward its final configuration for the proof-of-concept demonstration.

The team at Sandia National Laboratories had been designated to monitor the results, providing an independent verification of findings and outcomes. Simultaneously, a specialized team at the University of Arizona was tasked with preparing the necessary electronics and configuring ten units, referred to as "black boxes," which would house the energy generator.

To maintain absolute secrecy, the UA team was informed that their work was part of a critical component for an upcoming Mars mission, requiring immediate completion to meet a rigorous flight schedule. This explanation fit seamlessly into the university's culture of handling high-priority, high-stakes projects.

These "urgent" initiatives were commonly referred to as assignments for the UA "bicycle shop," a nickname for the department renowned for tackling technically demanding tasks under intense time pressure. This cover story ensured that no one outside the loop would find the project suspicious, allowing the work to proceed without raising unnecessary attention.

As the jet touched down at Davis-Monthan Air Base in Tucson, the next phase of their journey began. A convoy of armored Suburban SUVs awaited them on the tarmac, flanked by security personnel. The transition was seamless, with agents guiding them to the far outskirts of the base toward two waiting FBI helicopters.

As the SUVs approached, Bella squinted, her eyes narrowing in recognition. "I know him," she murmured.

Jack turned to her, curious. "Who?"

Her brow furrowed as she tried to place the familiar figure. As the SUV came to a halt, her expression shifted to one of astonished joy. "Small world," she whispered, her voice tinged with emotion. Without waiting, she stepped out of the vehicle and approached the man with purposeful strides.

"Gandhi!" she exclaimed, her voice breaking with warmth. She embraced him tightly, her gratitude evident. "It's so good to see you. Thank you for everything—for being there when I needed you most."

The man returned her hug, his demeanor calm but touched. "Bella, it's always an honor. And I'm sorry for your loss. Payman was a remarkable man. His legacy lives on in you."

Jack, now standing beside Bella, extended a hand. "You must be Gandhi," he said, his tone filled with respect. "Bella's told me so much about you—how you risked everything to save Payman. Thank you."

Gandhi shook Jack's hand firmly, offering a slight smile. Gandhi then offered a respectful salute, "The honor is mine, sir. Thank you for your service. I've seen your Service Record, very impressive, a 'pirate hunter'. You deserve a medal for bravery on the high seas". Gandhi chuckled and gave a slight bow of his head. "And congratulations to you both.

I understand a wedding is in order? Bella, you've always been remarkable, and it seems you've found someone equally special." He glanced at Bella with a knowing look. "Jack, you've found a good one."

Jack blushed slightly. "I'm the lucky one," he replied earnestly. Gandhi who was the commander of the USAF Guardian Angels team, is now retired and has transitioned to FBI Agent in Charge of HRT air operations.

With introductions and pleasantries exchanged, Gandhi ushered them toward the waiting helicopter. "Okay, let's get flying. I've been fully briefed, and I will be leading your security team."

Their handbags were loaded by the crew, and they buckled in. Once their headsets were on, they were briefed on the short flight ahead and what to expect on the mountain.

The helicopter ride to Mt. Lemmon was swift but significant. As they ascended, the sprawling Tucson valley came into view, a patchwork of desert beauty framed by jagged mountains. Bella stared out the window, her thoughts a mix of nostalgia and anticipation. Jack leaned closer, pointing toward the horizon. "See that ridge? That's home."

Bella nodded, her gaze softening. "It's beautiful. Remote, secure… perfect."

As the helicopter descended onto the property, the gravity of their mission settled over them. The secluded estate, nestled amidst dense pines, felt like a fortress—both a sanctuary and a launching pad for their ambitions. As the rotors slowed, Jack turned to Bella, his expression resolute. "This is it. Our new beginning."

With Henry bounding ahead, they stepped onto the grounds of their new home, ready to face the challenges ahead. The journey from Bentley Hall had been bittersweet, but here, amidst the quiet strength of the mountains, they found a renewed sense of purpose.

Their work, their love, their shared vision—it all pointed toward a future filled with possibility.

Bella smiled, taking his hand. "Together, Jack. Always together."

CHAPTER 54
RESULTS!

The Mt. Lemmon Research Lab

Catalina National Forest, Tucson, Arizona

12 January 13:33 Local Time

After weeks of relentless work and countless sleepless nights, the breakthrough finally arrived. The underground lab, nestled securely within Jack's estate, had become a haven of focused energy and determination. Their latest tests consistently proved that the Zeliconite crystalline material could reliably initiate and sustain cold fusion reactions in the modified black boxes. This was no longer a fleeting success—this was a reproducible reality.

The comfort of the estate, with its serene surroundings and meticulously maintained interior, stood in stark contrast to the intense, high-stakes work being conducted within its hidden depths. Despite the gravity of their work, the estate's warmth offered fleeting moments of solace—open windows letting in the crisp scent of pine, the gentle rustle of leaves outside, and the occasional bark of Henry darting through the grounds.

The lab, buried deep beneath the estate, was a sanctuary of cutting-edge technology. Within its walls, screens glowed softly, instruments hummed in rhythm, and the team buzzed with tireless energy. Jack, his eyes weary yet triumphant, scanned the final set of data. His voice, hoarse from lack of sleep, broke the silence. "We did it!" he said, the weight of his words sinking into the room. "The batches are stable. Every single one produces the same levels of energy. The black boxes work! This isn't just a dream anymore—it's a reality. You've all played a part in changing the world."

A moment of stunned silence followed, as if the team needed time to absorb what had just been confirmed. Then, an eruption of cheers filled the room. Scientists, engineers, and specialists who had poured their hearts and minds into the project shared hugs, high-fives, and handshakes. Bella, unable to contain her joy, threw her arms around Jack, both of them teetering on the edge of exhaustion but buoyed by sheer adrenaline. Their eyes, though rimmed with dark circles, gleamed with triumph.

Around them, champagne corks popped, and glasses were carefully filled. "To the future!" one scientist called, raising his glass. Another added, "To the team that made the impossible possible!" The toasts continued, each clink of glass carrying a weight of pride and accomplishment. Yet, despite the celebratory atmosphere, the team handled the moment with care—mindful of the delicate equipment surrounding them, the SCIF's sterile environment referring to the Sensitive Compartmentalized Information Facility (SCIF), and the enormity of their achievement.

"It's incredible, Jack," Marcy the communications team leader said, her voice filled with awe. "To see it work—repeatedly, consistently… This is going to change everything. The potential to power entire cities, to drastically cut greenhouse gas emissions—it's a dream we've turned into reality. We will be ready for the world 'GO LIVE' event!"

The team echoed the "GO LIVE, GO LIVE, GO LIVE" chant, unlike their normal reserved nature.

There were smiles all around, but then reality set in. Bella met Jack's eyes, her concern evident. "Jack, you've done it. Be careful what you wish for. This is your dream, but others may see it as either an opportunity or a curse."

Jack met her gaze, the magnitude of their achievement mirrored in his expression. "I get it. I understand. It's more than a dream," he said softly. "It's a responsibility." Leaning in, he hugged her, their connection a quiet acknowledgment of the journey they had undertaken together. For a moment, the world outside the lab could wait. In that embrace, they celebrated not just the success of their work but the partnership that had brought them here.

The embrace broke, Jack raised his glass for a final toast. "Thank you," Jack murmured, his voice breaking with emotion. Bella smiled, her hand resting gently on his cheek. "We've done it, team. But now comes the hard part. This discovery—it's a gift, but it's also a responsibility. We have to manage this carefully. The rollout, the partnerships, the patents—every step must be deliberate. And secrecy remains paramount."

As the hum of celebration settled, Michael, the head of security stepped forward. His voice, steady and firm, cut through the room. "This stays within the SCIF," he said, referring to the Sensitive Compartmentalized Information Facility. "What you've achieved here is groundbreaking, but it also makes us a target. The implications of this are massive—an energy revolution, clean and sustainable. But we can't afford any leaks. Every step forward must be secured. I don't need to remind us of the situations on UA campus or at Bentley Hall."

The gravity of his words settled over the team, sobering the atmosphere. It was an important reminder. Yet, even within that somber moment, there was an undeniable sense of pride. Together, they had achieved what many had deemed impossible.

Later that evening, as the estate quieted and the team dispersed, Jack and Bella ventured outside. Henry darted ahead, his small frame weaving between the pines with unbridled energy. The night air was crisp and cool, carrying the faint, comforting scent of the forest. For the first time in weeks, they felt the weight of their work lift, even if only temporarily.

Bella paused, gazing at the horizon where the estate's silhouette merged with the stars. "This place," she murmured, "has given us so much. It's not just a lab or a home—it's where we've made history. Funny to think, someday people will visit this site as a museum like Thomas Edison's Menlo Park."

Jack slipped his arm around her shoulders, drawing her closer and with a bit of humor, "Right, that's foretelling, we may never have our results known like the work of Edison. But like the prolific Edison, maybe this is just the beginning," he said softly. "What we've started here—this is going to change the world. Maybe the world will never know who reshaped the global environment."

They continued walking in silence, letting the forest's quiet embrace soothe their frayed nerves. The sprawling estate, a blend of rustic charm and modern innovation, felt like a sanctuary—one that had nurtured their dreams and now stood as a testament to their success.

With their success confirmed, Jack and Bella moved forward not only as pioneers of a new scientific frontier but as custodians of a technology that promised a brighter, cleaner future. The black boxes—compact, efficient, and transformative—were the culmination of their relentless effort. But they were also the beginning of a much larger journey. The team was back together in the SCIF.

Jack had the floor. "We need to engage with global energy leaders and policymakers from the start. This isn't just about the science. It's about introducing this technology responsibly, making it accessible and safe for everyone. Remember there are others; big energy business, world power brokers, and nations will want us stopped!"

Bella nodded, adding, "And we have to prepare for the inevitable resistance. We aren't the first to claim cold fusion, but we'll be the last. We need answers to every tough question, and we need to build trust—early and consistently."

The forewarning cast a chilling shadow over the team, its ominous weight pressing heavily on their every thought.

The team split into two focused groups to tackle the dual challenges ahead. Jack led the technical team, refining the black box design for efficiency and safety. They conducted rigorous simulations, testing the devices under extreme conditions to ensure their reliability. Meanwhile, Bella headed the communications team, crafting a comprehensive strategy to introduce the technology to the world.

In the weeks that followed, both teams worked tirelessly, their efforts synchronizing seamlessly. Meetings became a regular cadence, updates flowing between the groups to ensure alignment.

Jack's team perfected the black box units, while Bella's team prepared detailed informational packets, press releases, and private briefings for global leaders. The GO LIVE event.

By the time the teams reconvened for a final review, the momentum was undeniable. "We're there," Jack announced, his voice steady with conviction. "The black box units are consistent, efficient, and safe. I think we're ready."

"We have a comprehensive rollout plan," Bella added, her tone confident. "Security, public engagement, scientific transparency—it's all covered. The world is ready for this."

As they exchanged a long, anticipatory look, one question lingered in the air: Was the world truly ready for what they were about to unleash?

The invitation to Camp David came as a stunning, unexpected development for Jack, Bella, and their key team members. The chance to meet with President Janis Wynn wasn't just a high honor—it was a powerful affirmation of the national and global significance of their work. Yet, the invitation also carried an undercurrent of urgency, as though the weight of what they had discovered had already begun to ripple through the corridors of power.

Bella read the official letter twice, her hand trembling slightly as she set it down on the desk. She turned to Jack, who was already pacing in their study, his hands in his pockets, deep in thought.

"This isn't just about presenting our findings anymore," Bella said, her voice tinged with both excitement and unease. "Meeting at Camp David, with the President—it means this is no longer just science. It's diplomacy. It's politics. Our work is about to step onto the world's most public stage."

Jack paused, leaning against the desk as if the words carried the weight of a tectonic shift. "It sure is," he said quietly. "It's an honor—but also a huge responsibility. Bella, we need to be ready for anything this could lead to. What if they bypass the usual channels? What if they prioritize national interests over peer review? What if they shut us down?" His voice carried a note of apprehension, his gaze distant.

Bella reached out, placing a hand on his arm. "Jack," she said softly, "we've come this far together. We'll handle whatever comes next, just like we always have. Focus on the facts and the science. We'll show them not just that this works—but why it matters."

Despite Bella's steady reassurance, tension hung thick in the air as the team gathered to discuss this sudden shift. It was a contingency Marcy's communications team had planned for but never truly expected. Excitement clashed with apprehension as they speculated on what this meeting could mean. The words "Camp David" carried an almost mythic weight, as if their work was now part of something far greater than they could have anticipated.

"Let's not kid ourselves," Marcy said during the briefing. "This isn't just a demonstration—it's a geopolitical chess move. Every word we say, every detail we present, will be dissected and interpreted in ways we can't control. We have to be precise. Unshakable. Ready for questions we haven't even thought of yet."

Jack nodded, his jaw tight. "And it's not just about the science. It's about trust. If we get this wrong—even slightly—this entire breakthrough could be dismissed or, worse, weaponized. We're not just presenting a discovery. We're introducing a paradigm shift. The stakes couldn't be higher."

As preparations for the demonstration intensified, Commander Gandhi and his security team meticulously coordinated the transport of one of the black box units The device, cradled in reinforced containers and wrapped in layers of secrecy, was accompanied by a team of technicians and security personnel. Its journey mirrored the gravity of the task ahead—a piece of technology so revolutionary it demanded the highest levels of care and caution.

Meanwhile, Jack and Bella prepared for their own journey, their emotions a swirling mix of pride and trepidation. Despite reassurances about a casual dress code, they both felt the pressure to embody the professionalism and gravitas their work demanded. Jack went through his wardrobe like a man searching for armor, trying on three different combinations of suits and ties before settling on a gray single-breasted jacket, complimentary trousers, and a crisp white shirt with muted floral tie that Bella complimented with a soft smile. But Jack couldn't resist, he packed his black turtleneck with deep blue jeans.

Bella, equally deliberate, chose a long black skirt and a crisp white blouse, pairing it with a thick coat to ward off the anticipated January chill. As she adjusted her cufflinks in the mirror, she caught Jack watching her, his expression a blend of admiration and nervous energy. He smiled.

"Do you think we've covered everything?" Jack asked, breaking the silence. His voice was laced with uncertainty. "This isn't just about our research anymore. It's about how we handle the spotlight on a stage we've never been on before."

Bella turned to face him, her gaze steady and reassuring. "We'll manage, Jack. We always do. This isn't just about us—it's about the science. The facts. We've done the work. Now we need to trust it. We know the scientific stage, now the geo-political stage."

Even as she spoke, a flicker of doubt crossed her mind. The stakes felt impossibly high. The chance to directly influence national and international energy strategies was exhilarating, but the weight of responsibility pressed heavily on her chest.

As they packed their bags in the quiet of their room, their conversation circled back to the potential ramifications of the meeting. Would they be able to convey the full scope of their discovery's potential? Could they persuade the most powerful leaders in the world to see the profound opportunities it offered—and the dangers of mishandling it?

"This could change everything," Jack said, his voice almost a whisper. "Not just energy. Politics. Economics. The way the world functions."

"And that's why we have to get it right," Bella replied, her voice firm. "If we falter, if we let fear or pressure cloud our judgment, we risk losing control of what we've built."

Their resolve solidified as they zipped their suitcases shut. They were about to step into a role few scientists ever experienced—directly engaging with global leaders, not just as innovators but as stewards of a technology that could redefine the future.

As they boarded the plane the next day, Bella glanced out the window at the rolling mountains of Arizona fading into the horizon. The next chapter of their journey was about to begin, and the anticipation of what lay ahead crackled like electricity in the air.

They were ready—or as ready as anyone could be—to take the first steps onto the global stage, carrying with them nct just a discovery, but the hope and weight of a world waiting for change.

CHAPTER 55
CAMP DAVID

Andrews Air Force Base/Camp David Presidential Retreat Frederick County, Maryland
10 February Noon Local Time

The FBI Citation jet touched down at Joint Base Andrews, the weight of their impending disclosure settled heavier on Jack and Bella with every passing moment. The journey had been smooth and swift, but the proximity to their historic meeting added an undeniable edge. Waiting for them on the tarmac, once again, was Gandhi, their stalwart protector, whose steady presence provided a comforting constant in the chaos.

He greeted them with his trademark warm smile. "Welcome to Andrews. I trust the flight was comfortable? Just one more short hop to Camp David now."

Bella returned his smile, her expression softening. "It's always so reassuring to see you, 'St. Michael.'"

Gandhi chuckled, the kind of laugh that carried both levity and gravitas. "And how are the two of you holding up?"

Bella responded with a wry grin. "The flight was smoother than we could have hoped, thanks. But let's just say Jack is contemplating a fainting couch."

Jack gave a weak chuckle, clearly trying to mask his nerves. "A bit anxious," he admitted, adjusting his tie for the third time in five minutes.

"Meeting with the President to discuss something this transformative—it's surreal," Bella added, her tone more composed, though the tension in her fingers belied her calm demeanor.

Gandhi gestured toward the waiting SUV. "Well, no fainting on my watch. Let's get you there."

The vehicle, a subtly armored but comfortable SUV, carried them through the short drive toward Camp David. The dense forest surroundings offered a stark contrast to the arid Arizona landscape they had left behind. Jack stared out the window, his fingers drumming nervously on his thigh.

"Everything's moving so fast, Bella," he said, his voice quieter now. "Just a few months ago, we were in the lab troubleshooting prototypes. And now…"

Bella reached over, her fingers intertwining with his. Her voice was calm, a balm to his nerves.

"I know it's overwhelming, but think about why we're here. Our work has the power to change the world for the better. That's worth every bit of this whirlwind."

As the wooded scenery gave way to the secured gates of Camp David, the gravity of their mission settled further into their hearts. The retreat, steeped in history, had seen pivotal moments that shaped nations. Now, it would serve as the setting for another such moment—one that carried promises of global transformation.

Upon arrival, security personnel briefed them with military precision before leading them to a cozy cabin to freshen up. The air was crisp, the natural tranquility of the retreat a stark juxtaposition to the monumental discussion awaiting them. Each step felt heavier with expectation.

Gandhi, walking them to the cabin door, turned and offered a reassuring nod. "The President is eager to hear directly from you. She's been briefed, but your insights and passion will make all the difference."

Bella nodded, her gratitude genuine. "We're honored to have this chance. It's not just about the science—it's about what this discovery means for the world: energy independence, addressing poverty, stopping environmental collapse… Jack and I keep saying, maybe peace could actually break out."

Jack gave her an incredulous side glance. "Careful, you're starting to sound idealistic."

"I'm allowed to be, just this once," she shot back with a grin.

Inside the cabin, they sank into a well-worn leather sofa, the tension in their shoulders momentarily easing. Bella glanced at the door, where outside Gandhi stood guard, a silent sentinel. His presence, both as a protector and a friend, grounded her in the moment.

Jack pulled the presentation from his bag, but before they dove back into preparation, Bella rested her arm around his shoulder. "Hey," she said softly, tilting her head to catch his gaze. "We're almost there. You have this."

He exhaled a deep breath, his hand finding hers. "I know. I just keep thinking—what if we're not ready?"

Her grip tightened, her voice unwavering. "We are ready. Together, we've overcome every challenge so far. This isn't just about us—it's about everything this discovery can mean for the world. We'll face it, Jack, just like we always have."

Their clasped hands remained steady as they exchanged a glance that spoke volumes—a shared understanding of the road they had traveled and the uncharted path ahead. It was a moment suspended between hope and trepidation, between the weight of responsibility and the promise of extraordinary change.

CHAPTER 56
THE SUMMIT

Camp David Presidential Retreat

Frederick County, Maryland

10 February 13:30 Local Time

Jack and Bella, feeling the weight and significance of their impending meeting, opted to walk to Laurel Lodge rather than take the more expedient golf cart. The decision to walk gave them a last moment to collect their thoughts and absorb the serene environment of Camp David—a place steeped in historical decisions and diplomatic summits.

As they walked across the beautifully maintained grounds of the retreat, they passed Aspen Lodge, perched on a hill with a commanding view of the lush Maryland countryside. The lodge, historically used by every president since Franklin D. Roosevelt, stood as a symbol of continuity and the gravity of presidential authority. Reflecting aloud, Jack said, "It's incredible to think about the discussions that have taken place here—the decisions that have shaped history."

Bella gazed at the scenery. "And now here we are, walking the same paths as so many historic leaders, about to discuss how our discovery could shape the future of the world. It's both humbling and exhilarating."

They approached Laurel Lodge, the main cabin where meals were served and meetings were conducted. This was where they would present their research and discuss its potential to revolutionize energy consumption and its implications for national and global policy.

Stopping at the front door, they stood silently, composing themselves. Bella broke the silence, her voice steady but soft. "Whatever happens in there, we're in this together."

Jack didn't reply immediately, but the way he tightened his hold on her hand spoke volumes. Finally, with a deep breath, he said, "We just need to stay focused and clear."

They opened the door and were greeted by aides and security personnel, who directed them to the sitting area. The lodge's interior was warm and inviting, with a glowing fire in the large fireplace. The room felt less formal than a White House conference room but no less serious.

As they waited to be introduced into the meeting in progress, Jack and Bella reviewed their notes one last time. Before they realized it, the door opened, and they were invited to step inside. Announced to the conference room, Bella's heart skipped a beat as her eyes immediately locked on President Janis Wynn. The President, exuding confidence and authority, greeted them with a firm handshake and a commanding gaze.

"Drs. Arensico and Douglas, thank you for coming. I've been looking forward to our discussion. May I call you Bella and Jack, use your first names?" President Wynn said, her words carrying both warmth and gravity. Without missing a beat, she gestured toward the table. "Let's get started, shall we?"

"Yes of course fist names work." Responded Bella.

Taking their seats, Jack and Bella quickly realized this wasn't just a bilateral meeting. It was a multinational gathering, with significant figures such as the Prime Minister of the United Kingdom, Mr. Jennings, and key representatives from both the UK and US governments. The presence of these officials underscored the international interest in, and implications of, their cold fusion discovery.

"Jack, Bella, thank you again for joining us today," President Wynn began, offering a reassuring smile. "We have leaders and experts from both sides of the Atlantic, from both your home countries, who are eager to understand the potential of your work and discuss how we can collaborate on this groundbreaking initiative."

Prime Minister Jennings leaned forward, nodding. "Indeed, your discovery has the potential to be a cornerstone in the future of global energy and human policy. Can you briefly remind us of the importance of cold fusion?"

Jack took the question, "Briefly, cold fusion is a theoretical process where nuclear fusion—combining atomic nuclei to release energy—occurs at or near room temperature. This contrasts with traditional nuclear fusion, which requires extremely high temperatures, like those found in the sun.

Room temperature cold fusion will provide a nearly limitless, clean energy source without the harmful radiation or environmental impact of current energy technologies."

PM Jennings acknowledged, "Well done Jack, that was a simple definition. I was prepared for a scientist's lecture. We are simple politicians, not scientists and you have treated us with respect. We are here to ensure you receive the support you need and that your cold fusion implementation can benefit as many people as possible—in an equitable manner."

Bella gathered her thoughts, aware of the immense opportunities and responsibilities before her. "Thank you, President Wynn, Prime Minister Jennings," she began. "We appreciate the recognition of the potential our research holds. We're here to present not just the scientific basis of our discovery but also to discuss its practical implications and the strategic considerations for its deployment. We have prepared an extensive strategic communication plan, which we are ready to roll out to ensure a successful introduction."

Jack followed, his voice steady but passionate, as he delved into the technical aspects. He wove into the discussion the mechanics of their cold fusion technology, highlighting how it differed from previous attempts and why their approach had succeeded where others had failed. "Our process involves a unique crystalline material, Zeliconite," Jack explained, "that facilitates consistent neutron production, which is the key to stable cold fusion reactions. The scalability of this technology means it could be adapted for various uses—from small-scale local energy grids to major industrial applications—without the environmental degradation associated with traditional energy sources. We are prepared to take the world stage, of course, with your support."

The discussion soon shifted to the broader implications of introducing such a technology into global energy markets. Topics included the necessary regulatory frameworks, the potential for economic disruption in traditional energy sectors, opposition from oil, gas, and coal industries, and the geopolitical ramifications of a shift toward widespread use of cold fusion.

President Wynn focused the conversation on the immediate global balance of power. "It's crucial that we consider the security aspects as well. This technology could redefine global power structures and energy independence. We need a cooperative international approach to manage this transition. This will change how countries and factions relate to one another. The balance of power will be disrupted."

"Absolutely," Prime Minister Jennings added. "And we must also consider the social impact, ensuring that the benefits of this technology are accessible to all—not just economically advantaged nations or individuals."

The meeting continued with discussions on collaboration frameworks, funding mechanisms for further development, and strategies for public communication to manage expectations and promote accurate understanding of the technology's potential.

Jack and Bella listened attentively as President Wynn articulated the stance of the US and UK governments. "It is imperative that we manage the rollout of this technology carefully," the President said. "The global energy market, equity markets, international relations, and even national security could be destabilized if this is not handled correctly. The transition to this new energy source must be gradual. It needs a well-thought-out strategy that involves all stakeholders and aligns with global sustainability goals without precipitating economic turmoil."

Bella nodded thoughtfully, her attention riveted by the precision and clarity of the President's words. Under the table, Jack gave her hand a reassuring squeeze.

"We understand the concerns," Bella replied. "Our intention has always been to introduce our discovery in a way that benefits humanity while safeguarding environmental, economic, and social stability. We are prepared to work closely with both governments and the international scientific community to ensure this transition is as smooth as possible."

Jack added, "We appreciate the need for secrecy at this stage. We're committed to advancing this technology with the utmost responsibility. Our team is ready to follow your lead on how best to proceed with both the development and the eventual disclosure."

President Wynn smiled, perhaps impressed by their composure. "Thank you both for your sensitivity and professionalism. Your vision is exactly why this project has come this far. We'll set up a joint task force, including members of your team and our top advisors, to oversee the next phases. This group will refine the implementation plan, address security issues, and prepare for potential geopolitical shifts."

Prime Minister Jennings chimed in, "This task force will also consider how to involve other nations and international bodies. It's important that when the time comes, we present a united front that demonstrates thoughtful leadership and collaboration. However, this discovery must remain closely held for now. The world is a tinderbox, and if rogue nations or special interests acquire this technology prematurely, the consequences could be catastrophic."

As the meeting drew to a close, Jack and Bella felt the enormity of their roles settle on their shoulders. They had transitioned from researchers to pivotal players in shaping global policy—a daunting yet exhilarating prospect.

CHAPTER 57
FIREFIGHT

Camp David Presidential Retreat

Frederick County, Maryland

10 February 17:11 Local Time

The key players moved to the dining room for a casual yet elegant dinner served on presidential china. The warmth of the setting, with its soft lighting and quiet clinking of silverware, contrasted sharply with the undercurrent of tension Jack and Bella felt. Shortly after the main course was served, an advisor to the president stepped in and whispered in her ear. Her calm expression hardened subtly as she nodded and stood.

"We have a developing situation in Arizona. We're going to the situation room to monitor the activities," President Wynn announced, her voice steady but unmistakably urgent.

Jack and Bella exchanged glances, their concern mirrored in each other's eyes. Jack mouthed the word, "Arizona?" as the weight of the implication sank in. Without hesitation, they followed President Wynn and Prime Minister Jennings into the SCIF. As the heavy doors closed behind them, the gravity of the moment hit them like a tidal wave.

The situation room was a hive of controlled chaos. The hum of computers, the murmur of low conversations, and the faint crackle of radio chatter filled the space. Screens covering three walls displayed live feeds of the ongoing assault. Satellite images showed smoke rising from parts of the lab, while ground and aerial drone footage captured the chaos unleashed by the attackers. On the peripheral screens, other command centers patched in, their personnel equally focused.

Gandhi, already seated with a headset and directly connected to the HRT security team on the ground, barely looked up as he relayed updates. His usual composure carried a sharp edge now, his voice clipped and urgent.

President Wynn, exuding the calm authority of a seasoned leader, addressed the room. "We have a situation at your cold fusion facility in Arizona. It appears to be a coordinated attack by a well-equipped force."

Gandhi glanced at her and nodded. "Yes, Madam President, it was quite a surprise. I know the team leader well; he's prepared for every contingency. But an attack like this—broad daylight on American soil—wasn't something anyone expected. Madam President, I request posse comitatus be invoked to allow fighters from Davis-Monthan Air Force Base to engage. The Guardian Angels unit is ready to deploy; they're warming up choppers as we speak and awaiting orders."

President Wynn's eyes narrowed as she considered his request. She knew the legal weight of posse comitatus. "Prepare the authorization," she said decisively, directing her staff. "Notify the Attorney General to draft a retroactive presidential authorization to suspend Posse Comitatus and document this decision. We'll act now and justify later."

Jack and Bella stood frozen, their eyes glued to the live feeds. The lab, the place they had poured their lives into, was now a battlefield. The attackers seemed to have planned meticulously, targeting the critical research areas while sparing their home. It was a chilling realization—they knew exactly where to strike.

"Jack," Bella whispered, her voice barely audible. "They knew… they knew where to hit."

He nodded, his jaw tightening. "They've been watching us. Waiting."

The satellite feed flickered, shifting to drone footage of the attackers breaching the outer perimeter. Smoke billowed as explosions rocked the lab. The screen panned across the chaos—security personnel engaging the attackers, some pinned down while others fought valiantly to hold their ground.

"The initial assault took out our perimeter defenses," Gandhi reported, his voice sharp and direct. "Local security engaged, but we've lost contact with some personnel. Additional forces from Davis-Monthan are en route, and local authorities have been instructed to stay clear. The attackers are armed far beyond the capabilities of local law enforcement. The Guardian Angels choppers are now twelve minutes out."

Bella stepped closer to the screens, her gaze fixed on the layout of the lab. "What about the staff? Are they accounted for? And Henry?" she asked, her voice trembling slightly as she mentioned their beloved pug.

Gandhi glanced up. "The black boxes and Henry are secured in the vault with two security team members," he assured her. "The vault has an escape route through the mining tunnels if the situation warrants it."

Bella exhaled audibly, relief washing over her. The room's attention briefly turned to her, but she squared her shoulders, her composure returning. "We can't afford to lose any of the research—or the people protecting it."

President Wynn, her expression unflinching, turned to Gandhi. "We need to manage this crisis with precision. The potential fallout from losing control of this technology could have global repercussions. We will not lose this battle."

Gandhi nodded. "Understood, Madam President. The priority is the safety of personnel and securing the critical assets. If our team can hold the line, the reinforcements will give us the upper hand."

As the drone camera zoomed in on two attackers, Gandhi's expression hardened. "Looks like Quds Force operatives," he muttered. "This is bigger than we thought."

Jack and Bella exchanged a glance. The mention of the Quds Force sent a chill down their spines reminding them of the battle at Bentley Hall. The stakes were higher than they'd imagined.

Jack, normally reserved in such situations, stepped forward. "We need to prepare for the worst. If the vault is breached, we can't let the black boxes fall into their hands.

We'll connect to the SCIF in the vault and give instructions for their destruction if necessary."

President Wynn nodded, gesturing for an aide to make the connection. As the feed from the vault came online, Bella's heart leaped at the sight of Henry wagging his entire rear end. She leaned toward the screen, her voice soft yet firm. "Henry, stay safe, okay? We'll see you soon."

Henry barked in response, his tail wagging furiously. The moment brought a brief, much-needed reprieve from the tension, a reminder of what they were fighting to protect.

As the counterassault engaged, the situation room became a flurry of activity. Jack and Bella were quickly briefed on emergency protocols, including remote data protection and safeguarding transported materials.

They provided detailed building schematics, pointing out potential vulnerabilities and suggesting defensive strategies.

"Focus on the lab's east wing," Bella instructed. "That's where the secondary servers are. If they're compromised, we lose all backup data."

The screens showed the Guardian Angels choppers arriving, their rotors kicking up dust as they prepared to engage. The tide of the battle began to shift as reinforcements poured in, their superior firepower evident.

As they watched the unfolding events, Jack and Bella felt the weight of their work like never before. The dangers of their discovery falling into the wrong hands had never been so starkly clear. They stood together, their hands clasped tightly, silently vowing to rebuild if necessary and continue their mission.

Bella whispered, her voice resolute, "We'll get through this, Jack. We have to. The world needs this."

Jack nodded, his grip on her hand tightening. "We will. No matter what it takes."

The situation room buzzed around them, a symphony of chaos and strategy, but for Jack and Bella, the moment crystallized their resolve. This was more than a battle for their research—it was a fight for the future they had dared to imagine.

CHAPTER 58
THE CAVALRY

Camp David Presidential Retreat
Frederick County, Maryland
10 February 19:03 Local Time

On the ground in Arizona, the firefight intensified with each side exchanging relentless volleys of gunfire. The echo of shots ricocheted off the narrow valley walls, mingling with the sharp clang of granite shards splintering from boulders where defenders had taken cover. The air was thick with smoke, acrid gunpowder, and the palpable tension of combat. Overhead, a drone buzzed incessantly, its cameras feeding live footage back to the command center. It was through this surveillance that confirmation came: the attackers were elite Iranian commandos, their faces identified and matched within seconds by sophisticated facial recognition software.

Gandhi stood slightly apart from the others, his headset delivering a constant stream of updates. Despite the chaos, he maintained a composed posture, his eyes betraying no emotion as they darted across the screens. He fixated on the drone footage of the commandos, absorbing critical information without a flicker of reaction—a testament to his years of combat experience and unflinching resolve under pressure.

President Wynn, observing Gandhi's subtle but telling body language, sensed an escalation. "Gandhi, report," she commanded, her voice sharp with authority.

"Madam President," Gandhi began, turning slightly to address her, "the commandos are a diversion. Intel indicates this assault is part of a broader strategy. Simultaneously, a cyber-attack is targeting our financial systems and national lab infrastructures. It's highly coordinated."

A chill swept through the situation room. The assault wasn't merely a physical attack; it was a multifront assault, both tangible and digital. President Wynn's jaw tightened, her resolve evident. "Prepare for a multi-front defense," she ordered. "End the Arizona situation now. I want options for immediate countermeasures on my desk in five minutes. Notify the Attorney General to prepare the necessary documentation for invoking Posse Comitatus retroactively. Gandhi, manage defensive and offensive responses without hesitation."

The screens displayed a harrowing scene. Smoke billowed from parts of the lab, and aerial footage captured the attackers' ruthless precision as they closed in. Gandhi issued rapid instructions to the ground teams, his voice steady even as the situation deteriorated. The commandos had breached the perimeter, and the defenders were pinned down, fighting valiantly but losing ground.

The thumping rotors of HH-60W Pave Hawk helicopters cut through the smoky air, signaling the arrival of reinforcements from Davis-Monthan Air Force Base. Among them were the elite Guardian Angels, their arrival a lifeline to the beleaguered defenders. As the rotor blades' roar intensified, it merged with the relentless gunfire in a cacophony of hope and conflict.

The defenders, a mix of seasoned operatives and Guardian Angels, braced for the reinforcements. They understand all too well the gravity of their predicament. Meanwhile, drones armed with precision munitions circled above, and F-35 fighters provided high-altitude cover, ensuring no escape for the attackers.

Amid the chaos, an RPG slammed into the blast-proof exterior doors of the lab, followed by two more explosions in quick succession. Granite splinters and debris rained down, and the forest filled with the acrid tang of explosives. The doors withstood the attack. The overwhelming smell was as sharp as a finely honed KA BAR combat blade and just as threatening. The defenders were outgunned, their resolve tested against an enemy prepared to die for their mission. Yet, they held firm, determined to protect the critical research and their comrades.

Back in the situation room, the atmosphere was electric with urgency. The screens showed the Iranian commandos moving with military precision, their relentless assault testing every defensive measure. Bella's voice cut through the tension, her concern evident. "What about the staff? Is everyone accounted for? And Henry, is he still okay?" Her plea reminded the room of the stakes beyond the battle—a beloved pug, symbolic of home and humanity amidst the chaos.

Gandhi glanced at Bella, offering a sliver of reassurance. "The black boxes and Henry are secure in the vault, along with two team members. The vault has an escape route through the mining tunnels if needed."

Bella exhaled audibly, a flicker of relief softening her features. The room, however, remained charged with controlled chaos as Gandhi directed the counteroffensive.

On the mountain, the rhythmic staccato of the helicopter gunners' GAU-21 .50 caliber machine guns erupted, their rounds churning the earth and suppressing enemy positions. Two additional tactical helicopters arrived, disgorging reinforcements via fast ropes into the fray. The medics, armed with intel on injured defenders, moved with precision, stabilizing the wounded and pulling them to safety under relentless fire.

The air force troops fanned out, taking positions and reinforcing the beleaguered defenders. The presence of fresh forces breathed new life into those who have been holding the line, their morale bolstered by the sight of their comrades arriving to share the burden of battle. They launched a counteroffensive, reclaiming ground with grit and determination. Under the cover of aerial support, the tide began to turn. For the first time, the attackers faltered, their momentum shattered by the defenders' resolve and reinforcements.

The commandos, disoriented by the ferocity of the counterattack, retreated under the onslaught. Hope surged among the defenders—not just survival, but victory felt within reach. This battle was far from over, but the defenders had proven their mettle, reclaiming their ground with unwavering courage.

As the video screens in the command center displayed the defenders rallying and coordinating with the incoming reinforcements, Gandhi's headset buzzed with urgent news. He turned his back to the chaos of the screens and the ongoing live dialog from the firefight, listening intently. The voice on the other end was terse and grave, delivering information that twisted the already knotted situation further.

Gandhi switched his comms to speaker mode, the operator's voice clipped. "We have confirmation on the identity of the commando leader," the intelligence officer reported. "It's Payman Merhi ."

The name hit Gandhi like a physical blow, his brow furrowing deeply. "Payman? But intel confirmed he died last year," he said, disbelief clouding his normally composed tone.

"Apparently, those reports were mistaken," the officer replied, frustration laced in his voice. "He's very much alive and leading this assault."

The revelation sent a ripple through the command center. On the screens, the battlefield appeared to tip in the defenders' favor, with the remaining commandos retreating under heavy fire. Drone feeds streamed live footage of the carnage—bodies lay motionless, others staggered or crawled, their movements slow and defeated. Amidst the chaos, one feed zoomed in on a figure limping away. Bella froze, her heart pounding. It was unmistakably him. Payman.

"He's alive..." Bella's voice cracked with disbelief, the shock in her tone cutting through the tension in the room. Gandhi, standing beside her, stared at the screen, his expression hardening. "He didn't perish in the crevasse on the Swiss glacier," he murmured. "It was a ruse, a masterful deception."

A storm of emotions swelled within Bella—betrayal, anger, and a profound sense of loss. Rising abruptly from her seat, she tore the Rolex Sky-Dweller watch from her wrist—a gift she had chosen for Payman, a token of love that now felt like a cruel symbol of duplicity—and hurled it at the monitor as Payman moved across the terrain. The heavy watch struck the screen with a loud thud, distorting the image of Payman momentarily before destroying part of the LEDs. Bella's scream, raw and guttural, pierced the air. "You bastard! Lying, manipulative bastard!"

The sudden movement of anger triggered an instinctive reaction from a Secret Service agent, who stepped forward, his hand raised to intercept any further actions.

As the agent froze, assessing the situation, the atmosphere shifted. What had moments before been an air of collective alertness now softened into one of somber understanding. President Wynn moved to Bella's side, placing a comforting arm around her shoulders. Her voice, calm and empathetic, broke the tension. "I feel for you, young lady. Men can have their faults, especially those who play games like this."

Bella collapsed into Jack's arms, her sobs muffled against his chest as he wrapped her tightly in a protective embrace. Gandhi, ever the professional, resumed relaying information to the command center, his focus unwavering even as the emotional storm unfolded beside him.

Prime Minister Jennings leaned down, picking up the discarded watch. Turned it over in his hand, then handed it to Jack with a wry smile. "This looks valuable. Sturdy at least. You might want to save it for a better day."

Jack hesitated, caught off guard by Jennings' attempt to inject levity. "Right… uh, good money was spent on it." His awkward tone was met with a chuckle from the Prime Minister.

"I've often said," Jennings continued, his voice carrying a reflective weight, "if more women were in charge, we might avoid such brutal wars. Perhaps it's time to hand over the reins. I'll let one have my job." He finished with an audible chuckle.

The comment drew a faint smile from Bella, though her anger still simmered beneath the surface. Jack gave her hand a reassuring squeeze, his silent support grounding her.

Meanwhile, the dual command centers remained tense and focused. Analysts and military strategists murmured among themselves, their voices blending with the constant hum of monitors and the rapid clatter of keyboards. Each center worked in synchronized precision, their tasks split between managing the cyber-attack and the firefight on the mountain.

Cheers erupted from the cyber team as Gandhi confirmed the reason, the cyber threat had been neutralized. Their focus shifted to retaliatory measures, ensuring the attackers paid a steep price for their brazen actions. On another screen, live feeds from the mountain showed the commandos retreating to the 'City of Rocks,' a labyrinth of house-sized granite boulders spread across the mountain. The defenders, aided by drones and infrared scans, pursued the attackers methodically, clearing the area stone by stone.

Despite their best efforts, Payman eluded capture. Drone footage revealed no additional heat signatures, and the search teams recovered seven Iranian commando bodies. Two wounded assailants were taken into custody for interrogation. The defenders' losses, however, were deeply felt. Two HRT members had been killed, and four others were airlifted to Banner University Medical Center at University of Arizona a Level 1 Trauma Center for medical care. The remaining forces worked to fortify defenses, ensuring no further breaches.

In the situation room, a collective breath was finally drawn. The immediate threats had been neutralized, and the cyber-attack decisively quashed. Yet the weight of the day's events hung heavily over everyone present. Bella, still seething, leaned into Jack's arms, her anger giving way to exhaustion. The mention of Payman's escape reignited her fury, but for now, she allowed herself to rest, drawing strength from Jack's unwavering presence.

As the room settled into an uneasy calm, President Wynn addressed the assembled advisors. "Today was a reminder of the stakes we're dealing with. This technology doesn't just represent hope; it's a flashpoint for conflict. We must remain vigilant, unified, and ready for whatever comes next."

Bella, her voice quieter but no less resolute, added, "We'll rebuild, stronger than before. They can't break what we've built, not if we stand together."

Around her, nods of agreement rippled through the room. The fight wasn't over—not by a long shot. But in that moment, amidst the tension and triumph, there was a renewed sense of purpose. They weren't just defending technology; they were defending a vision for a better, safer future. And for Bella, Jack, and everyone in that room, the battle was only beginning.

CHAPTER 59
ECHOES OF DECEPTION

Camp David Presidential Retreat
Frederick County, Maryland
12 February 08:00 Local Time

The early morning sun barely crested the horizon over Camp David, casting a soft glow over the secluded retreat that had housed days of intense deliberations and emotions. POTUS had returned to the duties of the White House. Inside, the air was thick with the last vestiges of tension that clung even as the final meeting adjourned. The deal was struck—a calculated division of power masked by diplomacy. Ten black boxes, the core of an advanced capabilities system, were to be split evenly. Five would remain under the tight security of the US Atomic Energy Commission (AEC), while the remaining five would remain in the high-security confines of the Mt. Lemon lab, under the watchful eyes of Bella and Jack and a reinforced security team. AEC team members would be embedded with the team to provide technical support and security. Research would continue on the peaceful applications of the groundbreaking technology. The lab and retreat would be hardened.

As the politicians and their entourages dispersed, Bella found herself outside, breathing in the crisp morning air, trying to shake off the lingering claustrophobia of paneled rooms and frenzied actions. Her phone buzzed with a reminder of the scheduled video call from Mt. Lemon, checking in.

Bella moved from the tranquil embrace of Camp David's woods into the stark, fluorescent-lit SCIF. The room hummed with urgency; monitors flickered with real-time data, and the soft clatter of keyboards mixed with the low murmur of intense discussions. It was a sharp contrast to the stillness outside, a reminder of the gravity of their mission and the very busy SCIF. The science and technical teams conferred via video teleconference as they worked final details of their return to Arizona. Jack was busily engaging in a dialog about expanding the lab to accommodate the increased focus on Zeliconite development.

Suddenly, Bella in a cheerful voice, called out, "Hi Henry, my good boy" as Henry responded with a vocal response of his own, but what really gave away his excitement at hearing Bella was the way he wagged the entire second half of his body, wagging with joy.

Michael, their head of security replied, "Henry will not leave the lab on his own, we must encourage him with "treats" to move to the house or to go outside. If we're working through the night he's here with us if not and we get him to the house, he's on your bed. He's waiting for you Bella."

Bella smiled, imagining his warm, comforting presence. "We're almost through this, boy, we'll be home soon" she murmured, a reassurance for both Henry and her.

Henry continued wagging his tail vigorously at the sound of her voice. Each time she spoke Henry's ears perked up, his head twisted and eyes searched the monitor screen as if he could sense her. He longed for her return home.

Michael continued the dialog about getting the lab and home in order. "The lab is 80% there we do have final configuration of the equipment we just received from Sandia. But I know, I've said this before and I'm happy to repeat myself, there was no, zero damage to the lab. We were locked in the vault in the mine shaft. All is well. The work topside is progressing, and your home received minimal damage except to some glass and stray bullet holes. But it will be 100% by the time you return."

Meanwhile up above the lab, even in the dark of the Arizona morning, the remnants of the Arizona firefight were being meticulously erased. The home that had served as a covert base was now being restored to its immaculate facade. Workers moved quietly and efficiently, each aware of the gravity. Their silence held. Every bullet casing, every sign of struggle was wiped clean, leaving behind no echo of the violence that had transpired.

That same morning, POTUS addressed the nation. Her tone was somber yet resolute as she described a tale of cyber warfare—a bold narrative of Iranian forces attacking the US energy infrastructure, a crisis averted by the nation's robust cyber defenses. "We remain vigilant and ready to respond at a time and manner of our choosing," she declared, the script crafted to reassure a public unaware of the actual skirmish that had unfolded on domestic soil.

No mention was made of the battle that had raged in the quiet mountains of Arizona. Instead, a carefully worded press release painted a different picture: one of heroic intervention by the USAF Guardian Angels who had thwarted a cartel kidnapping attempt. It was a narrative neatly woven into the fabric of ongoing national concerns of the US, Mexican border, cartel kidnappings, and drug threats, bolstering support for increased funding and advanced technology to fortify the porous Tucson corridor.

As the day wore on, Bella found herself reflecting on the events that had led them here. The deception was necessary, they told her—a shield to protect not just national security but also the very fabric of public trust. Yet, each omission, each fabricated story weighed heavily on her. She wondered about the cost of such secrets, the invisible burdens they carried, and whether the truth, when distilled through layers of necessity and strategy, lost its essence along the way.

As evening enveloped Camp David, the tranquil solitude offered Bella a momentary respite from the whirlwind of meetings and decisions that had consumed the last several days. Wrapped in her Camp David logoed winter jacket she sat on a secluded bench overlooking the darkening woods. The crisp air pinched her cheeks. She allowed herself the space to reflect on the magnitude of what they have undertaken.

Her role placed her at the nexus of a clandestine alliance, one that was now guardian to the world-altering secret of cold fusion. This breakthrough, capable of redefining the global energy landscape, was mired in the complexities of geopolitical strategies and deep-rooted secrecy. The weight of this knowledge was immense, and the responsibility of safeguarding it even more so.

The recent firefight at her home and lab in the mountains flashed through her mind. A reminder of the stark reality of the dangers they faced. The attack had been repelled, the evidence erased, but the scars remained—etched not only on the land but within her. She smiled as she reflected on the safe outcome at the lab and of Henry waiting to sniff and greet her.

The stakes extended far beyond her own safety or even the immediate bounds of national security. Cold fusion was a key to rewriting humanity's future—a chance to end energy scarcity and reshape economies. Yet, Bella couldn't ignore the haunting specter of its misuse, the chaos it might unleash if it fell into the wrong hands.

Turning to Jack as they sipped their tea, she murmured, "Do you ever think about how fragile this balance is? A single misstep, and everything we've worked for could become a weapon."

Jack nodded, his face somber. "Every day," he said quietly. With that thought, Jack rose and indicated that he was returning to the cabin. He suggested that she take some time with her thoughts. She should return when she was ready.

This technology could end energy scarcity, reshape economies, and redefine geopolitical alliances. Yet, these possibilities were shackled by the necessity of secrecy, driven by the fear of its potential misuse or the chaos it might unleash if revealed prematurely. The aggressive actions of the Iranian commandos proved that there was already a threat and more to come.

As the evening deepened, the soft rustle of leaves in the gentle wind seemed to whisper change. Bella pondered the path ahead. How long could they maintain this facade? What would be the consequences when the world finally learns of cold fusion? Her role was clear, yet the ethical dimensions cast long shadows.

Her thoughts were interrupted by a distant call of an owl, a solitary sentinel in the night. Bella looked up, her gaze piercing the darkness, her resolve strengthening, as the harbinger of wisdom she carried for the future. This secret, this power, was a double-edged sword—one she must wield with great care.

In her heart, Bella knew the road ahead was fraught with challenges and moral quandaries. But tonight, under the cloak of darkness and a blanket of stars, she reaffirmed her commitment to steer this course with integrity, knowing that the dawn would eventually bring light not just to the skies but to the shadows of deception that clouded their mission.

She stood to return to their cabin, the cold-crisp evening air felt like the calm before a storm—a storm she was now central in navigating. As Bella walked back to the cabin she shared with Jack, the crunch of gravel underfoot was rhythmic, grounding her amidst the swirl of her thoughts. When she reached the door, it creaked softly as she pushed it open, by now a familiar sound a gentle herald of her return. Inside, the cabin was dimly lit, the soft glow of a single lamp casting warm shadows across the wooden walls. Jack turned, his warm smile breaking through the tension like a shaft of light. "Long day?" he asked gently. Her heart held a mix of emotions—weariness from the day's burdens and a comforting anticipation of the solace that Jack's presence always provided.

Jack folded some documents they had reviewed earlier. Bella was pleased when she saw he was now wearing the Rolex Sky Dweller. After her self-described tantrum when she saw Payman, she asked him to accept it with her love. He of course did so. It was no longer a reminder of her loss; it was a reminder of the bond of their love and the emotion of Payman's duplicitousness she jettisoned. The sight of Jack, strong and steadfast, was a balm to the tumult of her thoughts.

He held out a steaming mug of tea, one with an aroma of chamomile. His face broke into a smile as his eyes met hers. It was a smile that spoke of understanding and shared secrets, a smile that carried the weight of their joint burdens with a grace that only love could offer. Without a word, she crossed the room and enveloped him in her arms. The tension that Bella had carried all day melted away in the strength of his embrace.

Their connection was a bond forged in fire and secrecy. Jack held her, the layers of duplicity and strategy that cloaked their days began to peel away, leaving just the raw, unguarded truth of their shared humanity.

They moved together to the small comfortable living area. They carried their teacups. A teapot sat on the table, the herbal aroma filling the space with a soothing scent. Together they sat sipping quietly, each lost in their own thoughts yet comforted by the other's presence.

"Last night here," Jack murmured, his voice a soft rumble against her ear. "Tomorrow, back to the real world."

Bella nodded, her head resting against his chest. "The real world," she repeated, the words tasting bittersweet on her lips. Here, in the isolation of Camp David, cloaked in the urgency and shadows of national security, a microcosm governed by necessity and tight-lipped silence. But beyond the gates, reality awaited, with all its challenges and unyielding demands.

Preparing for bed, the cabin felt like a sanctuary, a temporary shelter not just from the physical elements but from the storm of their responsibilities. In each other's company, the world outside—with its relentless pace and unending complexities—seemed momentarily distant.

Tonight, they would find rest in each other's arms, a respite that fortified them for what was to come. Bella, as she drifted off to sleep, felt a profound gratitude for Jack's presence in her life, his strength not just a shield but a source of profound personal comfort. Tomorrow would bring its trials, but tonight, they had each other, and that was enough.

CHAPTER 60
FURY AND FORTITUDE

Arizona Quantum Lab Mt. Lemmon
Tucson, Arizona   01 April

Now back for a couple of months to the mountain lab and retreat, work on the next phase of their strategy took precedence. The team had fallen into a routine of continuous confirmation of results under various conditions. All findings were very promising. The next steps focused on rolling out cold fusion to the world. While they were immersed in their work, evil forces were at work elsewhere.

World news was now focused on the aggression of Russian attacks on the Ukrainian people and their electrical grid. These attacks were relentless. The tyrant Russian president, a dictator, had made a strategic error in underestimating the resolve of the Ukrainian people, their leaders, and their worldwide allies. What was supposed to be a week-long military exercise of Russia rolling into Kyiv had turned into a bloodbath for both sides and had dragged on for years. For the Russians, the loss of lives was staggering. They emptied their jails, conscripting criminals and forever altering the lives of a generation of healthy young men who had once looked forward to domestic life with their girlfriends and wives. And were now maimed or dead. Russia had depleted its army. The young, fighting-age men were plunged into the meat grinder of battle. Meanwhile, the majority of educated and talented young men had emigrated by any means possible to avoid what many believed to be the idiotic pipe dream of the madman at the helm of their once-great country.

The resolve of the Ukrainian people in the face of the Russian invasion had been notably strong and unified, showcasing a remarkable spirit of resistance and resilience. This determination was embodied in various symbolic acts of defiance and unity across the nation—and even within Russia itself. One poignant example was the viral commemoration of a Ukrainian POW's last words, "Glory to Ukraine," which sparked widespread admiration and tribute within hours on social media. Murals were painted, and discussions of erecting a monument in his honor began.

In the dim light of a candlelit shelter in Lviv, a young Ukrainian volunteer whispered to his comrades, "They want us to break, but every blackout just makes us light more candles." His voice carried both defiance and hope, embodying the spirit of a nation that refused to surrender.

Ukrainian morale had been a significant factor in the ongoing conflict, contradicting initial expectations of a quick collapse in the face of the invasion. Instead, the Ukrainian public, including tens of thousands of volunteers and those enlisted in the military, united in a formidable display of national unity against the aggression.

This unity had not only sustained Ukrainian resistance but also demoralized parts of the Russian military forces, as evidenced by reports of appeals and protests from mobilized Russian troops. Russia's mercenary army, Wagner, had deserted and marched on

Moscow in protest of Russia's inability to supply their fighters on the front. That action was swiftly quashed, and their leader was assassinated on orders from the Russian dictator.

The skies over Kyiv often turned a smoky gray, a harbinger of missile strikes that left neighborhoods in ruins. This collective resolve and the unyielding spirit of the Ukrainian people, especially the youth, had been central to the nation's resilience in the face of adversity, shaping a strong, unified identity centered around resistance to the invasion and the pursuit of a sovereign, democratic future.

Russian strategists, determined to weaken the resolve of the brave Ukrainian people, focused on targeting civilian infrastructure. These actions raised significant humanitarian and legal concerns. Since October 2022 and continuing, Russia had launched multiple waves of missile and drone attacks against Ukraine's energy infrastructure, leading to widespread blackouts and damage to the power grid. These attacks had not only disrupted daily life for millions of Ukrainians but had also posed grave threats to critical sectors such as healthcare and municipal services. This emphasized the urgent need for emergency power supplies like generators and fuel to keep essential facilities operational.

Legal experts and international bodies scrutinized these actions under the lens of international law, particularly the Law of Armed Conflict and the Geneva Conventions, which set forth rules for the protection of civilian populations during armed conflicts. Attacks causing excessive civilian casualties or damage to civilian objects without direct military advantage were considered violations of these laws, potentially constituting war crimes. While Russia claimed that its strikes on energy infrastructure were aimed at weakening Ukraine's military capabilities, the disproportionate impact on civilians and civilian infrastructure raised serious questions about the legality of these actions under international humanitarian law. The Russian dictator was clearly a war criminal.

The head of the U.N.'s atomic watchdog agency condemned a drone strike on one of six nuclear reactors at the Zaporizhzhia Nuclear Power Plant in Ukraine, saying such attacks "significantly increase the risk of a major nuclear accident." Officials at the plant acknowledged that the site was attacked by drones, including a strike on the dome of the plant's sixth power unit. The International Atomic Energy Authority confirmed physical impacts from the drone attacks at the plant, including one of its six reactors.

As the sun set over the charred remains of another bombed-out Ukrainian town, the world's collective anger simmered like a pot on the brink of boiling over. Images of children huddled in makeshift shelters and elderly survivors sifting through rubble continued to flash across global screens, igniting outrage in the hearts of millions. Nations, once divided by political and ideological barriers, found common ground in their condemnation of Russia's brutal and unrelenting aggression.

The anger was not just directed at the dictator orchestrating the invasion but at the unyielding tide of destruction and suffering his decisions had unleashed. The world watched with clenched fists as the Ukrainian people, unbowed and unbroken, stood as a symbol of defiance in the face of tyranny. Their resilience became a rallying cry for freedom-loving nations, a beacon of hope amidst the darkness of war.

Leaders across the globe issued stark warnings, their voices laced with frustration and determination. "Russia's actions have not only violated international law but the very essence of humanity," one European leader declared. Sanctions tightened like a noose around the Russian economy, but still, the dictator pressed on, his madness pushing the world closer to a tipping point.

In the quiet corners of the free world, ordinary citizens channeled their anger into action—protests erupted in streets, funds were raised for humanitarian aid, and prayers were whispered in countless languages for a swift end to the carnage. Yet the tension remained palpable, a global community waiting, holding its breath, as the storm of conflict continued to rage.

The world's fury, like the conflict itself, was far from contained. It surged and swelled, a force that could no longer be ignored. The question now loomed heavy over every decision, every strategic move: How far would the world go to stop the madness? And how long before this fury transformed into a united, unrelenting force to end the dictator's reign of terror once and for all? Or the dictator cornered and unleashing the devil of the atom, nuclear weapons.

CHAPTER 61
ON TO THE FUTURE

Arizona Quantum Lab Mt. Lemmon
Tucson, Arizona 02 April 06:42 Local Time

On a beautiful spring morning, with birdsong greeting the day, Bella woke to the morning sun rising over the Catalina Mountains. The open windows allowed the fresh scent of the pines to envelop her, mingling with the warmth of the early rays. As she lay in bed, slowly emerging from the last remnants of sleep, another aroma caught her attention—the rich, inviting scent of freshly brewed coffee. Bella had to admit she was a coffeeholic, and this was enough to energize her. She smiled, already anticipating the comforting bitterness of the first sip.

The absence of Jack in bed should have been her first clue, but she was still surprised when he appeared, greeting her with a kiss, a hug, and a steaming mug.

"Coffee and kisses for my love," Jack said warmly, his affection as palpable as the aroma of the coffee.

Bella grinned and accepted the mug. "You know how to start my day perfectly."

"Come out to the balcony and enjoy the morning with me," he said, his voice full of tenderness.

The balcony offered a commanding 180° view of the valley below, their home perched at a 4,500-foot elevation. It was an especially clear day, the kind where the horizon seemed endless. Bella took her favorite seat, savoring the sweeping view as she cradled her mug. Jack leaned against the railing, his gaze distant.

"I love this view," Bella said, breaking the silence. "I can see all the way to Mexico and everything in between."

"Uh-huh," Jack replied absentmindedly, his expression distant. Bella turned to him, raising an eyebrow. "What does that 'uh-huh' mean? You're not with me."

Jack sighed, pulling his phone from his pocket. "I've been reading the news. The Russians are still holding the Rivne nuclear plant hostage, withholding vital energy from the region. They've just launched new attacks on Ukraine's energy infrastructure, destroying 80% of eastern Ukraine's grid. It's like they're making one final push to bring the country to its knees."

Bella's face darkened. "That Russian dictator pisses me off. You know there's something we can do, Jack. We have a solution to Ukraine's energy crisis."

Jack shook his head, taking a deep breath. "We do, but don't forget we signed an agreement with POTUS. We're forbidden from sharing our findings with anyone else, ever. I mean, maybe not under penalty of death, but pretty damn close. It's all about keeping the world safe."

He hesitated before continuing. "You know how badly they want our knowledge. Those terrorists sent Payman here, along with a team of Iranian Quds Force commandos, to kill us and take it. They wouldn't have made such a bold move if they didn't think we had something. Payman was their ace, and they played him. They'll be back—we just don't know when or how. Keeping this 'genie in the bottle' is the only way to stay ahead."

Bella sat forward, her tone sharp. "We've proven that we're smarter and tougher than they are. And we have the full force of U.S. Special Forces backing us up. We can't just sit here and let that tyrant roll over Ukraine and then move on to the Baltics. He has no concern for anyone—just his power and money."

Jack lowered himself into the chair next to her, his head in his hands. The vibrant morning seemed muted by the weight of their discussion. The weight of their responsibility bore down on Jack, every decision feeling like a thread in a web that could unravel at any moment. He sipped his coffee slowly, the silence stretching between them like a taut wire. Finally, he looked up.

"You're right. You're always right. I can't argue against that. But now we need to figure out how to do this without letting the genie out and without breaking our declaration."

Bella's eyes lit up. "I have an idea. Do you remember when we were in the situation room with the president? She slipped me her personal cell number and said, 'If you ever need anything—anything—call me directly.' I think this qualifies."

Jack laughed dryly, shaking his head. "You really think she'll answer? That was probably just a courtesy. I like her, I trust her—but let's face it. She's juggling Ukraine, domestic terrorists, and white nationalist crazies trying to turn this country into their own personal dystopia. How does she have time for us?"

"Let's find out," Bella said, standing decisively. "Let's go down to the SCIF and call her number. See what happens."

Jack hesitated, then nodded. "Alright. What do we have to lose? At least we'll be doing something instead of waiting for someone else to fix things. I'm all in."

Bella's excitement was infectious, and soon Jack found himself propelled by her energy as they made their way through the house and descended into the SCIF.

The secure facility, hidden behind reinforced doors and buried deep within the old mine, was cold and damp despite the steel and concrete that had transformed it into a fortified stronghold. The amber lighting cast long shadows, softening the space but failing to erase the chill. The cold bit through their clothes, a stark reminder of the isolation their work demanded.

Bella dialed the number on speakerphone, her fingers trembling slightly. Instead of a normal ringtone, a series of electronic beeps and zings filled the air, reminiscent of an old dial-up internet connection. She exchanged a wide-eyed glance with Jack—both thinking the same unspoken "What the hell?"

Suddenly, the line went silent. Just as Jack reached for the end button, a voice cut through the quiet.

"Hello, Dr. Arensico. This is Secret Service Agent McCourt on the president's personal line. This is a secure connection. I've verified the signal is coming from a scrambled line. If you're in danger, may I have your code and sitrep?"

Jack mouthed a silent "Wow," his eyebrows raised.

Bella cleared her throat, her voice steady. "No immediate danger, Agent McCourt, but we believe we can impact the Ukraine crisis significantly."

There was a pause before McCourt responded, his tone professional but devoid of emotion. "The president is currently leading a cabinet meeting. Since this is not an emergency, may I arrange for her to return your call in twenty-two minutes?"

Bella glanced at Jack, who nodded. "Of course. Thank you," she replied, ending the call with a smile.

Jack leaned back in his chair, his skepticism returning. "I don't think we'll hear from her. Maybe an aide, but not the president."

Bella smiled, her determination unwavering. "We'll see."

As they patiently waited, Bella broke the silence. "I could use a cup of coffee." Together, they walked to the single-serving Nespresso Verturo. Bella got first dibs, choosing her favorite rich, dark frothy blend. Jack followed with a cappuccino.

"So, we wait," Bella said, settling into a chair. The two sat silently, sipping their coffee, the tension between them only slightly eased by the warmth of their mugs.

Exactly 22 minutes after their call with Agent McCourt, the phone rang. Bella almost aggressively punched the button. The line came to life with the familiar voice of Janis Wynn, POTUS.

"Hello, Bella. So, I've been told you have a solution to world peace. It's always nice talking with you. Do you have a new superpower invention?" POTUS's tone carried a mix of humor and curiosity.

"Well, I certainly think so, and thank you for calling, Madam President. Jack is here with me," Bella replied, glancing at Jack.

"Hello, Jack. I'm kind of rushed—what's on your mind?" Wynn asked, her voice more serious now.

"Madam President," Jack began, "I'm so fed up with these men—these autocratic dictators—creating wars to prove their manhood. I've seen what Putin has done to Ukraine's power grid. Our fusion units are small, portable, and capable of powering entire cities. I know we've signed a compartmentalized secrecy agreement, but Bella and I would like to offer one of our units to President Zelenskyy to replace the energy supplies they've lost while they repair their infrastructure and fight off Russia."

Jack's words hung in the air, and Bella nodded in agreement, her eyes locked on the speakerphone. The agreement they had signed with the U.S. government flashed through her mind—the five units they were allowed to keep for experiments, their guarantee of security, housing, and income for life. She knew they were walking a tightrope.

President Wynn's voice broke the silence. "Well, Jack, I never imagined this from our call. In keeping with our nondisclosure, I can tell you we have additional units in storage for emergencies. I like your thinking, but we need to keep this quiet. No one outside a select few knows about your invention, and I intend to keep it that way. Let me think about this. In the meantime, I want you and Bella on a jet immediately to meet with my advisors. A Citation jet is waiting for you at Davis-Monthan Air Force Base. Agent McCourt will send an escort to pick you up. Be ready in thirty minutes. Leave your units in place. Do not bring them. You'll be briefed further when we next speak. I have to go, but I like your idea. Safe travels."

The call ended, and Jack and Bella exchanged a look of determination.

There was a knock at the armored door of the SCIF, followed by the voice of Agent Berriman over the intercom. "I don't know how you do it, but we just received a directive from POTUS to get you to DM for a wheels-up flight in 58 minutes. Grab your go-bags. We're leaving now."

Jack responded quickly. "10-4, we're moving."

As the security entourage loaded the armored Suburban, the distant rhythm of helicopter blades filled the air. Minutes later, an HH-60W Pave Hawk landed on the helipad, and Jack and Bella, along with Berriman and two bodyguards, boarded the aircraft. The flight to Davis-Monthan was short, and soon they were landing near a hangar on the outskirts of the base. Sure enough, a white business jet awaited them, with Gandhi standing at the stairs to greet them.

The jet was luxuriously configured, with eight large reclining seats, table with its own chairs, and an aft bedroom. As the plane reached cruising altitude, the co-pilot entered the cabin to ensure their comfort and inform them of a classified flight plan to Kyiv. He also mentioned a scheduled call with POTUS in five minutes, to be taken in the aft cabin, which doubled as a secure communications area.

Jack and Bella exchanged glances, marveling at how quickly their morning had shifted from sipping coffee on their balcony to cruising at 45,000 feet on a mission to change the course of history. They shared a brief laugh to ease the tension.

When the call came in, they moved to the aft cabin and answered. POTUS's voice was clear and direct. "Hello, you two. I hope you're comfortable and not too surprised by this sudden mission. I can now tell you that the government has produced additional units based on your designs, all portable and ruggedized for any environment. Ten of these are en route to Ukraine. You will arrive three hours ahead of them. President Zelenskyy will be waiting for you. You will be my representatives on the ground. No one is to know the full extent of your technology. Anyone who comes into contact with the units will be sworn to secrecy, knowing that failure to comply could mean the loss of their country to an invading dictator. Are my orders clear?"

Jack began to respond, but Bella silenced him with a gentle touch and a smile. "Yes, Madam President. We understand and will carry out your directive."

"Thank you," Wynn said. "Safe travels. Make this work. I truly believe we can turn the tide of this war. God bless the USA and Ukraine." With that, the call ended.

As they sat back, the weight of their mission settled on them. Bella turned to Jack, her eyes sparkling with determination. "Let's spend the flight in bed. I've always wanted to join the mile-high club and experience making love while traveling at 600 mph. Go grab a bottle of champagne. I'll warm up the sheets."

Jack grinned, walking to the main cabin to retrieve the champagne. Gandhi and Berriman exchanged knowing smiles, reassured that their charges were in capable—and united—hands.

The plane soared eastward, carrying them toward a war-torn nation and a brave leader admired for his fortitude, Bella and Jack prepared to do their part in changing the course of history. The secret they carried would either save Ukraine or bring new dangers. For now, the world waited.

ACKNOWLEDGEMENTS

This book is a work of fiction, inspired by a lifetime of experiences and exposures of a full life. Most important in the development of this work, was the unforgettable summer journeys I spent hiking the Camino de Santiago and exploring England and Scotland by train. They say the pilgrimage of the Camino brings one closer to a higher spiritual plane, perhaps even closer to a higher power. For me, it was a journey enriched every day by the companionship of four remarkably bright, intelligent and inspiring women: Jennifer, Liliana, Gabrielle, and my incredible wife, Marcy. Each of you brought challenges, insights, and laughter that transformed the journey into an unforgettable experience. To my wife Marcy, my soul mate, and fellow adventurer thank you for being my constant source of encouragement and joy. Your companionship through every step of life and every mile of the Camino makes each journey richer I am forever grateful for your love and support

The science would not have been possible without the invaluable guidance of Stephen Fleming from the Center for Quantum Networking at the University of Arizona. His sage advice and expertise helped to shape my early ideas and continually sharpen my focus throughout the process. Steven, thank you for your insight, patience and dedication, which were instrumental in bringing this story to life.

To provide a solid foundation in quantum physics and mechanics, Marcy and I attended courses at Oxford University, continuing education program, where I had the privilege of meeting many brilliant and inspiring individuals. Among them was John Hind an Englishman of exceptional knowledge and talent, whose keen editorial insight was invaluable in refining my work. John skillfully sharpened my prose, correcting my Americanisms and ensured that the fictional scientific theories in the book were somewhat rooted in plausible reality. Beyond his education editorial brilliance, he was also a practiced punter and an extraordinary flamenco guitarist a testament to his diverse talents.

I am grateful to my friend, Dr. Burton Bentley, a board certified emergency physician, who introduced me to the complexities of the tib-fib injury - a severe and sometimes life threatening condition. His enduring friendship, shared memories, and remarkable ability to speak expertly on nearly any topic, provided inspiration well beyond his professional expertise. His historically resonant name even became the foundation for Bentley Hall in this story. Burton, thank you for your friendship for your ideas you've sparked along the way.

Two great patriots—Jolly Green pilots and Combat Search and Rescue (CSAR) leaders—helped bring this novel to life. Tim Healy, the real "Ghandi," warrior and helicopter industrialist, inspired the inclusion of CSAR's elite warriors through his stories and legacy. Some say Ghandi himself is responsible for the nickname for the HH-60W, the 'Whiskey.' This revised edition was further shaped by CSAR hero, a rare humble leader, Major General (Ret.) Tom Kunkel, whose insight sharpened the military dialogue throughout. As a CSAR warrior, his example honors the spirit that drives this story—That Others May Live.

Lisa Laughlin encouraged me, read my early drafts, maintained support, and held me to my commitment to finish.

For these people I am forever grateful.

Thank you for reading Quantum. I hope the journey with Bella and Jack meant as much to you as it did to me in creating it. If it resonated, a quick positive review helps more than you know. To follow what's next—and Bella and Jack's continuing story—visit quantumfusionbook.com. See you in the future.

— Mark

www.ingramcontent.com/pod-product-compliance
Lightning Source LLC
LaVergne TN
LVHW010648051026
849532LV00001B/210
* 9 7 9 8 3 4 9 6 0 9 1 9 0 *